D0900343

METHODOLOGICAL FOUNDATIONS
OF
RELATIVISTIC MECHANICS

METHODOLOGICAL FOUNDATIONS

OF

RELATIVISTIC

MECHANICS

MARSHALL SPECTOR

UNIVERSITY OF NOTRE DAME PRESS
NOTRE DAME — LONDON

Library of Congress Cataloging in Publication Data
Spector, Marshall, 1936–
    Methodological foundations of relativistic mechanics.

    Includes bibliographical references.
    1. Relativistic mechanics.   I. Title.
QA808.5.S6   530.1'2   72-3508
ISBN 0-268-00472-2 (cloth)
ISBN 0-268-00488-9 (paper)

Manufactured in the United States of America by
NAPCO Graphic Arts, Inc., New Berlin, Wisconsin

For
Nan, Tony, and Jessie

ACKNOWLEDGMENTS

A number of people offered suggestions for the improvement of this book. Among them were Gary Gutting, John Lango, Robert Metzger, Myron Wecker, and David Weissman. I hope they are not too displeased with me for not taking all of their advice.

For general encouragement and good counsel, in philosophy and otherwise, I would like to thank Edward Erwin and David Weissman, and my wife Nan Spector.

M.S.

Setauket, Long Island
September, 1972

CONTENTS

INTRODUCTION

At present most books on the special theory of relativity fall into one of two classes. First, there are physics texts on the subject. These usually presuppose a familiarity with classical mechanics and electrodynamics, and proceed with the mathematical elaboration of the theory. Little attention is paid to basic methodological and philosophic issues, for example, the general nature and purpose of theories of mechanics. Second, there are a large number of popularizations of relativity theory. These employ little or no mathematics, sometimes sacrifice accuracy for dramatic effect, and too often leave the reader with a distorted picture of the nature of the theory and its relation to the tradition of scientific investigation and theory construction from which it arose.

Books of either type may well serve the purposes for which they were written: the training of the professional physicist in the use of the theory, and the education and entertainment of the interested layman. But there is an important function that they are ill-suited to serve: the provision of a grounding in the theory for one whose primary interest is in the philosophy of science, so that he may read with understanding the sometimes technical literature on the philosophical implications of relativistic mechanics. One of my purposes in this book is to serve this function by providing a clear account of the methodological foundations of the theory, introducing the appropriate mathematical ideas as they become necessary or useful, but not relying solely on the mathematics to express the ideas under consideration.

In one sense, then, this book is an attempt to bridge the gap between the popularization and the physics text or treatise. But this is not the only or indeed the primary purpose of the book, for my motive in writing it is to analyze and clarify certain fundamental concepts, principles, and procedures in both classical and relativistic mechanics. I will pay close attention, therefore, to philosophic issues rather than merely developing theorem after theorem. This book, then, is an attempt at fundamental philosophic clarification embedded in the format of an introductory work. (This dual purpose will result in what might be deemed an unevenness of originality. Some sections of the book will be purely expository, and others will be polemical in the sense that I will analyze certain concepts and principles in a manner which disagrees with standard interpretations.)

The remainder of this introduction will be devoted to an outline of the presentation.

The first half of the book deals with the methodological foundations of classical mechanics, as conceived prior to the Einsteinian revolution. I begin by introducing the vector notions of position, velocity, and acceleration, with some emphasis on the role of a frame of reference in defining these concepts. The idea of force is then introduced as it was conceived by physicists working within the mechanical world view of the nineteenth century. Newton's laws of motion are presented, and some time is spent in developing the concepts of a mechanical force law, mechanical problem, mechanical explanation, mechanical parameter, mechanical system, and mechanical world view. This will amount to a detailed description of what Thomas S. Kuhn would refer to as the classical mechanics "paradigm"—a description of what distinguishes *mechanical* explanations, systems, etc. from other kinds of explanations, systems, etc. in physics.

It is shown, for example, how the resultant force acting on a body is determined by various parameters describing the situation in which the body finds itself: its position, velocity, and acceleration with respect to other bodies in its vicinity, the "nature" of the body and others in its vicinity, certain universal constants, and certain other coefficients depending in various ways on the specific problem at hand (for example, the stiffness coefficient in Hooke's law).

It will be useful, in order to give the reader a clear preliminary indication of the kind of discussion he will encounter in this book, to say a few more words about the classical mechanics paradigm before going on to outline the remainder of the book.

Two basic kinds of mechanical problem can be distinguished. First, there is the problem of determining the class of motions a system of bodies may undergo when the force law (or laws) are given. Solving a problem of this kind consists of combining Newton's second law of motion with the force law(s) appropriate to the system in question, and solving the resulting differential equations for the positions of the bodies as functions of time. Problems of this type are essentially exercises in applied mathematics (although not necessarily easy exercises). They are typical of problems found in mechanics textbooks, and in an obvious sense may be considered "trivial."

The second type of problem is: Given the motions of a system of bodies, determine the force law(s) responsible for these motions. This type of problem is more characteristic of certain kinds of basic research in physics. (There is a sense in which this kind of problem, or a generalization of it, is found in all branches of physics—not just mechanics.) But it turns out that this type of problem can always be solved in a trivial manner. For if the positions of the bodies in a system as functions of time are given, along with Newton's laws of motion, it is always possible to calculate back to *some* function of the parameters of the system which will both satisfy the second law of motion and yield the given motions of the bodies. (For one body, given Newton's second law, $\mathbf{F} = m \, d^2\mathbf{r}/dt^2$, and the observed motion of the body, $\mathbf{r}(t)$, one can always construct a function of $\mathbf{r}$ and the set P of other parameters of the system, $f(\mathbf{r}, P)$, such that $f(\mathbf{r}, P) = m \, d^2\mathbf{r}/dt^2$.) But of course not all such functions are considered as being *explanatory* of the given motions of the bodies of the system, though they may well be satisfactory from the point of view of the engineer.

Therefore, this second type of problem becomes physically interesting (nontrivial) only if we restrict the allowable kinds of force laws and the set of allowable parameters which may appear in these force laws. It is here that an element of *convention* as to what constitutes a mechanical explanation of the behavior of a system enters the picture. One attempts to *reduce* some force laws to others which are considered simpler, more basic, perhaps more "intelligible," or more purely "mechanical." (I shall also introduce and discuss a "parameter interpretability" condition.) It should be clear that an investigation of the kinds of implicitly imposed restrictions under which the physicist works would bring to light an important aspect of the classical mechanics paradigm.

All of the above is presented without presupposing a strong background in mathematics on the part of the reader, but with sufficient mathematical apparatus, properly introduced, to capture the spirit of what it is to have a "problem in mechanics," to have a "mechanical explanation," to have a specifically *mechanical* parameter, and so on.

The next stage in preparing the ground for the theory of relativity is a presentation of classical electrodynamics, once again paying close attention to basic methodological issues. I then discuss the senses in which electrodynamics can be conceived basically as a branch of mechanics, and how in fact it was so conceived for a time during the last half of the nineteenth century. This involves a general description and analysis of the attempt to actually reduce the electrodynamic parameters (the field vectors, for example) and the Lorentz force law to more "purely mechanical" parameters and force laws. This will amount to a description and analysis of just what was going on when physicists attempted to construct what have been called "mechanical models of the ether." Some light will also be shed on the general nature of model-thinking in physics, and the concept of the *reduction* of one theory to another—issues with which I have dealt to some extent elsewhere. (These attempts at reduction failed, resulting in one aspect of what Kuhn would call the "crisis" state of interlocking "anomalies" that developed toward the end of the nineteenth century.)

At this point the presentation goes back to certain first principles and certain assumptions that had been tacitly made earlier in the development. The concept of a *frame of reference* is examined once again, the notion of a change of reference frame is introduced, and the Galilean transformation equations are unearthed as the previously implicit "common-sense" rules for translating a description of a physical process or a physical law from one frame of reference to another. The notion of an inertial frame of reference is also uncovered, and the behavior of the laws of classical mechanics and electrodynamics under a Galilean transformation between inertial frames is discussed. The Michelson-Morley experiment is then presented as an important part of the crisis situation discussed earlier.

Against this background of a clear presentation of the methodological foundations of classical mechanics, the special theory of relativity can be introduced in such a way that it does not appear as a piece of exotic theory construction and argumentation that is disconnected with the tradition from which it arose. For now Einstein's ideas can be presented as a (revolutionary) response to the crisis situation which existed in the foundations of classical mechanics and electrodynamics at the turn of the century.

The special (or restricted) principle of relativity is presented, and it is shown how the Lorentz transformation equations can be derived if we assume the special principle and take certain experimental results (for example, the null result of the Michelson-Morley experiment) at face value. Having done this, it will appear that the laws of classical mechanics fail to comply with the special principle of relativity together with the Lorentz transformation equations as the mode of translating from one inertial frame to another. It is then shown how Einstein solved this problem with the bold step of introducing new mechanical laws (specifically, a replacement for Newton's second law of motion—a law which some philosophers and physicists had believed to be demonstrable *a priori*). In this way the special *theory* (as opposed to 'principle') of relativity can be introduced. That is, *relativistic mechanics* is presented and compared with classical or Newtonian mechanics.

It is at this point that the value of having spent a great deal of time in presenting the foundations of classical mechanics (and electrodynamics) will become apparent. For, in presenting relativistic mechanics much of what was said with regard to the methodological foundations of classical mechanics can be taken over without change. In this way the naturalness of Einstein's mechanics will stand out— something which rarely occurs in other presentations of the theory of relativity.

After completing the systematic presentation of the foundations of relativistic mechanics, I go on to discuss a consequence of the theory that has seemed to some to be of particular philosophic significance—the famous equation $E = mc^2$ and the related idea that mass and energy are in some sense convertible into one another.

In a concluding chapter I draw together and expand upon a few points that are treated only briefly earlier in the book.

A final word about the extent of my use of mathematics is in order. I shall assume, to put it concretely, that the reader has taken a course in calculus at some time in his educational career, and has all but forgotten what he learned. In the earlier chapters, therefore, I shall introduce concepts such as the derivative of a function as they are needed. The presentation is self-contained, but I hope that it will jog the reader's mind into recalling bits and pieces of what he once knew. Those who know calculus may find the presentation at this stage boring (at worst, I hope). When it comes to more difficult mathematical *derivations* later in the development, I shall ask the reader to "trust me" in the claim that something does indeed follow from something else; for what is of methodological importance is that he understand the physics that is being plugged into the mathematical machine, and the physics that comes out. He may then treat the actual derivation as a mathematical black box which experts certify to work without error—generating true statements when true statements are plugged in. Therefore, familiarity with the mathematics is not essential for understanding the ideas under consideration. I could have taken the course of using much less mathematics, but it seemed to me that inclusion of the appropriate mathematical contexts would make the book more useful to those who are able to follow the derivations.

1

CLASSICAL KINEMATICS

Generally the science of classical mechanics deals with the various types of motion that massive bodies undergo when subject to various types of forces.

I shall describe only *particle mechanics*; we shall deal only with "point masses"—systems consisting of massive bodies whose mutual distances are large in comparison with their sizes (or largest diameters) or in which for other possible reasons their sizes and shapes do not matter. (In Chapter 2 I shall indicate how systems in which this is not the case may be handled on the basis of particle mechanics.) To establish these notions clearly, let us look at the concept of *motion*.

POSITION

Briefly, motion is change of *position* with time. In order to give a determinate sense to these notions—to speak of different positions and different times—we need a *frame of reference* and an associated *coordinate system* with respect to which we can measure the position of a body at different times. Our frame of reference will be three mutually perpendicular rigid rods (indefinitely extendible) and a set of synchronized clocks. We associate a Cartesian coordinate system with this frame of reference by marking off equal distance intervals along the rods with a unit of length measure, and by marking off equal time intervals with a unit of time measure on the faces of the clocks. Following the usual convention, we label the rods the X, Y, and Z axes.

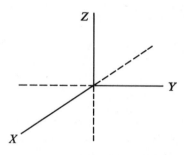

We could also refer to the set of clocks as the T axis, but this way of treating time on a par with the spatial coordinates becomes significant only when we come to relativistic mechanics. (A number of the concepts just introduced become problematic in relativistic mechanics. For the present, we take them as being clear enough.)

The *position* of a body at a given time t_1 (which we can label or refer to as '$P(t_1)$') is determined with respect to our frame of reference and its associated coordinate system as follows: We take the perpendicular projections of $P(t_1)$ along the three axes. The distances of these projections

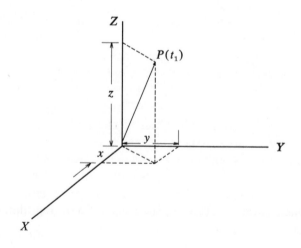

along the axes will be three numbers: $x\ (P(t_1))$, $y(P(t_1))$, $z\ (P(t_1))$; or more briefly, where confusion will not result, x, y, and z. This *ordered triple* of numbers, written as (x, y, z), can be looked upon as giving the position of the body at the given time. It is a "code" which we can use to locate the body.

We can also refer to the position by an arrow drawn from the origin of the coordinate system to the point P. The triple (x, y, z) then represents the three projections of the arrow along the axes of the frame of reference. The arrow can be referred to in boldface type as **r**, a *vector*, with x, y, and z being the vector's three components, labelled as r_x, r_y, and r_z.

We have, then, several equivalent ways of referring to, recording, or "encoding" the position of a body at a given time: $P(t_1)$; (or, briefly, P, where no confusion will result); the triple (x, y, z); **r** (or, obviously, **r** (t_1)); or the triple (r_x, r_y, r_z). Each of these notations has its own advantages and disadvantages in different contexts of discussion or problem-solving. It is useful to be actively aware of the possibility of different notations for the same ideas, so as to concentrate on "the ideas themselves" rather than their notational trappings. I shall go freely from one notation to another as convenience dictates in different contexts. This may slow down the reader at times, but it will force him to look beneath the notation.

Of course, we need not have used a Cartesian coordinate system as a basis for referring to, measuring, and recording the positions of bodies. We could have used, for example, one of the two coordinate systems shown in the figure below—each of which is associated with the same original *frame of reference*.

In the "spherical coordinate system" shown at the left we can refer to the position of the body as P (as before—since

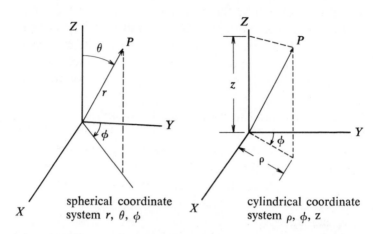

spherical coordinate
system r, θ, ϕ

cylindrical coordinate
system ρ, ϕ, z

this is just the *name* of the position and thus independent
of the kind of coordinate system used), or again by the boldface letter **r**, indicating an arrow drawn to P from the origin.
But now we take as the ordered triple of numbers (r, θ, ϕ),
where r is the length of **r**, and θ and ϕ are the angles shown
in the figure. In the cylindrical coordinate system shown at
the right the ordered triple (ρ, ϕ, z) is used, where each
member of the triple refers to the measure shown in the
figure. We have, then, several different "codes" that can be
used for recording the position of a body.

Which of these coordinate systems is best? There is no
general answer to this question. If one is working with a
system of moving bodies in which there is a large degree
of spherical symmetry, the use of a spherical coordinate
system will greatly simplify the expression of spatial relations among the bodies, as well as any necessary mathematical computations. Similar remarks apply to the cylindrical
coordinate system. There are many other kinds of coordinate systems which can be concocted, each of which may
be highly useful for certain kinds of systems of bodies in
motion and cumbersome for others. But each is equally
"correct." Each will allow one to uniquely determine the

position of a body.

Since we are not yet concerned with specific mechanical systems, I shall use a rectangular Cartesian coordinate system. Such a coordinate system is most useful in general discussions, because all three coordinates are, in an obvious sense, alike. (A Geocentrist might find spherical coordinates "more natural"; a Newtonian, conceiving space as a large box, might prefer rectangular Cartesian coordinates.)

VELOCITY

If a body is at position P_1 (or r_1) at time t_1 and at position P_2 (or r_2) at time t_2, it must have *moved* from P_1 to P_2 during the time interval. (We ignore the esoteric possibility that the body goes out of existence at P_1 and then reappears at P_2.) Its *average velocity* during the time interval is *defined* as

$$v_{av} = \frac{r_2 - r_1}{t_2 - t_1}$$

where the components (see p. 3) of the vector $r_2 - r_1$ are $x_2 - x_1$, $y_2 - y_1$, $z_2 - z_1$. That is, v_{av} can be thought of as a vector, represented by an arrow beginning at P_1 and ending at P_2. Its components are

$$\frac{x_2 - x_1}{t_2 - t_1}, \frac{y_2 - y_1}{t_2 - t_1}, \text{ and } \frac{z_2 - z_1}{t_2 - t_1}$$

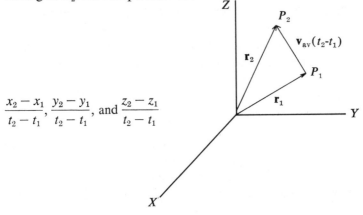

Each component records the average velocity along each coordinate axis. If we let '$\Delta\mathbf{r}$' (read as "delta r" or "change in r") be an abbreviation of '$\mathbf{r}_2 - \mathbf{r}_1$', and '$\Delta t$' be an abbreviation of '$t_2 - t_1$', we can rewrite the last equation as

$$\mathbf{v}_{av} = \frac{\Delta\mathbf{r}}{\Delta t}$$

This more perspicuous notation reflects the common-sense notion that average "speed" is equal to "distance" (this will be qualified below) traveled divided by the time required to traverse the distance.

There are, of course, various ways in which a body can get from P_1 at t_1 to P_2 at t_2. It may move along differently shaped trajectories (paths or routes), and it can move in different ways along any given trajectory—smoothly or in a "jerky" fashion, and so on. The average velocity vector $\mathbf{v}_{av}$ or triple

$$\left(\frac{\Delta r_x}{\Delta t}, \frac{\Delta r_y}{\Delta t}, \frac{\Delta r_z}{\Delta t} \right)$$

does not record these possible variations. It records only the straight line distance between P_1 and P_2 and the time elapsed (and, of course, which of P_1 and P_2 was the starting point).

Consider a body moving along the trajectory shown in the figure. The body is at position P_o at time t_o, and at P_1 at some later time t_1. Its average velocity during this time interval is, as before,

$$\mathbf{v}_{av} = \frac{\mathbf{r}_1 - \mathbf{r}_o}{t_1 - t_o} = \frac{\Delta\mathbf{r}}{\Delta t}$$

Now we may look upon P_1 as one of a sequence of positions $P_1, P_2, P_3, \ldots P_i \ldots$ closer and closer to P_o, and

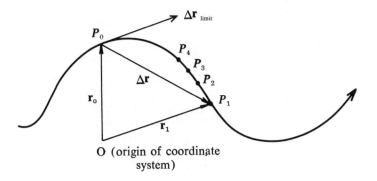

O (origin of coordinate
system)

we can consider the associated sequence of average velocities generated for each of these positions, taking P_o as the
"starting point" for each of these average velocities. For
each P_i in the sequence the corresponding Δr_i will be
smaller, but so will its corresponding time interval $t_i - t_o$.
The quotient $\Delta r_i / \Delta t$ will in general change for successive
values of i, but it will approach a *limiting value* (if the
motion is "not too jerky") as P_i gets closer to P_o. This
limiting value can be represented diagramatically by a
"Δr_{limit}" arrow drawn tangent to the trajectory at P_o. Using
these ideas, we can define the notion of *instantaneous veloc-
ity* of a body *at a position* as follows:

$$v(P_o) = \lim_{P_i \to P_o} \left(\frac{r_i - r_o}{t_i - t_o} \right)$$

where t_i is the time at which the body is at P_i.

Of course, as P_i approaches P_o, t_i approaches t_o, so we
can rewrite the definition of instantaneous velocity as

$$v(P_o) = \lim_{\Delta t \to o} \left(\frac{\Delta r_i}{\Delta t} \right)$$

If we have an equation for the trajectory giving the position
of the body as a function of time, $r(t)$, we recognize the

above expression as the *first derivative* of that function with respect to time, evaluated at P_o . That is,

$$\mathbf{v}(P_o) = \frac{d\mathbf{r}(t)}{dt}\bigg|\text{ evaluated at } t_o$$

$d\mathbf{r}/dt$ is itself a function, expressing the *rate of change* of the original function $\mathbf{r}(t)$ with respect to time. This *instantaneous velocity vector* $\mathbf{v}(P_o)$ also has three components—its projections along the three coordinate axes:

$$v_x(P_o) = \frac{dx}{dt}, \text{ etc. for } y \text{ and for } z$$

ACCELERATION

The instantaneous velocity of a body—the rate at which its position is changing with time—may itself change as it moves along its trajectory; that is, $\mathbf{v}(P_i)$ may have different values at different positions P_1, P_2, P_3, ... The body may slow down or speed up, or change the direction of its motion. (At the point of its reversal, P_r, we would have $\mathbf{v}(P_r) = 0$. This possibility was referred to earlier as "jerky" motion.)

A body whose instantaneous velocity is changing with time is said to be *accelerating*, and the concepts of *average* and *instantaneous* acceleration can be introduced in a manner analogous to the concepts of average and instantaneous velocity. Thus, *instantaneous acceleration* at a position is defined as follows:

$$a(P_o) = \lim_{P_i \to P_o} \left(\frac{\mathbf{v}(P_i) - \mathbf{v}(P_o)}{t_i - t_o} \right)$$

(This is the limit of a sequence of *average* accelerations.)

Using a procedure similar to that used above in the case of velocity, this can be rewritten in a clearer form:

$$a\,(P_o) = \lim_{\Delta t \to o} \left(\frac{\Delta v}{\Delta t} \right)$$

If we have an equation expressing v as a function of time, $v(t)$, we recognize the above expression as the first derivative of that function with respect to time, evaluated at P_o. That is,

$$a\,(P_o) = \left. \frac{d\,v\,(t)}{dt} \right|_{\text{evaluated at } t_o}$$

dv/dt is a function expressing the rate of change of the original function, $v(t)$, with respect to time. But $v(t)$ is the first derivative of $r(t)$, so we can write:

$$a\,(P_o) = \frac{d}{dt} \left(\frac{d\,r\,(t)}{dt} \right) = \left. \frac{d^2\,r\,(t)}{dt^2} \right|_{\text{evaluated at } P_o}$$

The instantaneous acceleration vector, **a**, is the *second derivative* of the position vector **r** with respect to time.

The instantaneous acceleration vector also has three components along the three coordinate axes:

$$a_x = \frac{d^2 x}{dt^2}, \text{ etc. for } y \text{ and for } z$$

In a similar manner we could define higher derivatives, but we will have no need of them in what follows.

A body is at *rest* during a time interval if for every instant of time t in that interval $v(t) = 0$. A body is said to be in *uniform translatory motion* during a time interval if for every instant t in that interval $a(t) = 0$; that is, if

$$a\,(t) = \frac{d\,v\,(t)}{dt} = 0 \text{ for all } t \text{ in the interval}$$

This will hold as long as $v(t)$ is equal to some *constant* vector during the interval. The body is then moving along a straight line at constant *speed*, where the speed of a body is the *magnitude* of the velocity vector, given by

$$\left| \mathbf{v} \right| = v = \sqrt{v_x{}^2 + v_y{}^2 + v_z{}^2}$$

—just as the distance of a body from the origin of the coordinate system is the magnitude of the position vector **r**, given by

$$\left| \mathbf{r} \right| = r = \sqrt{x^2 + y^2 + z^2}$$

We see that a body can move at constant *speed* (not velocity) even though its *acceleration* is nonzero if the *direction* of the motion is changing. An example of this possibility is uniform circular motion. Let us briefly consider this case.

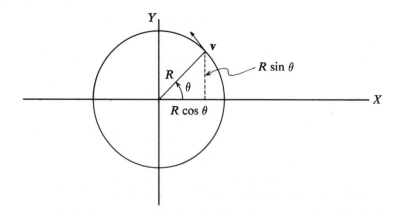

The body is moving, let us suppose, at constant *speed* in counterclockwise fashion as shown in the figure. That is, the angle θ is increasing according to the formula $\theta = \omega t$, where ω is a constant which is a measure of the "angular speed" of the body. The X and Y components of the position of the body can be read from the figure; they are

$$x = R \cos \theta = R \cos \omega t$$

$$y = R \sin \theta = R \sin \omega t$$

The components of the velocity vector **v** are

$$v_x = \frac{dx}{dt} = -\omega R \sin \omega t$$

$$v_y = \frac{dy}{dt} = \omega R \cos \omega t$$

The *speed* of the body, $|\mathbf{v}| = v$, is

$$v = \sqrt{v_x{}^2 + v_y{}^2} = \sqrt{(\omega^2 R^2 \sin^2 \omega t) + (\omega^2 R^2 \cos^2 \omega t)}$$

$$= \sqrt{\omega^2 R^2 (\sin^2 \omega t + \cos^2 \omega t)} = \omega R, \text{ a constant}$$

(The reader should check the form of this result against his intuitions about the case.)

The *speed* v is a constant; it does not change with time. But the acceleration is not equal to zero, for the body *is* accelerating: its *velocity* is *not* constant, as we now show.

The components of the acceleration vector **a** are

$$a_x = \frac{dv_x}{dt} = \frac{d}{dt}(-\omega R \sin \omega t) = -\omega^2 R \cos \omega t$$

$$a_y = \frac{dv_y}{dt} = \frac{d}{dt}(\omega R \cos \omega t) = -\omega^2 R \sin \omega t$$

The magnitude of the acceleration vector, $|\mathbf{a}| = a$, is given by

$$a = \sqrt{\left(\frac{dv_x}{dt}\right)^2 + \left(\frac{dv_y}{dt}\right)^2}$$

Performing the requisite calculations yields

$$a = \omega^2 R$$

Once again, a nonzero constant. This means that there *is* an acceleration; the velocity (though not the speed) is changing because the *direction* of the motion is continuously changing.

We have now completed a sketch of that part of classical mechanics known as *kinematics*. A full discussion of *classical kinematics* would also include a detailed study of various kinds of trajectories. But we shall rest content with this discussion of the concepts of position, velocity, and acceleration, and the basic concept of a frame of reference and its associated coordinate system. There is one more important kinematical concept whose discussion must be postponed until a later chapter—the concept of a change of reference frame (*not* coordinate system).

The difference between classical kinematics and classical *dynamics* (which will be presented in the following two chapters) is that the former considers the motions of bodies without regard to their *masses* or to the *forces* which are *responsible* for their motions. Kinematics, therefore, gives us the underlying framework within which the dynamics of bodies in motion can be presented.

2

CLASSICAL DYNAMICS
PART 1

We are now in a position to elaborate on the statement with which Chapter 1 began: classical mechanics—more specifically, classical *dynamics*—deals with the kinds of motions a system of massive bodies will undergo when subjected to the influence of different kinds of forces. (Once again I shall describe in detail only *particle dynamics*, where this is to be understood as in Chapter 1.)

We postulate the following law of nature as stating the effect of a force acting on a body:

When a force $\mathbf{F}$ acts upon a body of mass m at a time t, the body will experience an *instantaneous acceleration* $\mathbf{a}$ (see Chapter 1) due to that force given by

$$\mathbf{F}(t) = m\,\mathbf{a}(t)$$

From the considerations of Chapter 1 we see that this can also be written in the following equivalent forms:

$$\mathbf{F}(t) = m\frac{d\mathbf{v}}{dt} = m\frac{d^2\mathbf{r}}{dt^2}$$

All of the above expressions assume that the mass of the body does not vary with time—an apparently innocuous assumption. If, for whatever reason, we do not want to make this assumption, then the correct form of the law is:

$$\mathbf{F}(t) = \frac{d}{dt}(m\mathbf{v})$$

13

This will reduce to the other forms whenever

$$\frac{dm}{dt} = 0$$

since

$$\frac{d}{dt}(m\mathbf{v}) = m\frac{d\mathbf{v}}{dt} + \mathbf{v}\frac{dm}{dt}$$

I shall assume that the mass of any body under consideration does *not* vary with time, and will therefore restrict myself to the earlier forms of the law. (The more general form becomes theoretically significant only when we come to compare classical dynamics with relativistic dynamics in later chapters.)

The law we have assumed is called *Newton's second law of motion*.[1] We can write it in the canonical form of a law of nature familiar from the philosophical literature as follows:

$$(\xi)\,[(m\,(\xi) = M) \cdot (\mathbf{a}\,(\xi) = \mathbf{A}) \cdot (\mathbf{F}\,(\xi) = \mathbf{f}) \cdot \supset \cdot (\mathbf{f} = M\,\mathbf{A})]$$

where M is a constant, $\mathbf{A}$ and $\mathbf{f}$ are vector functions of several variables (to be discussed later), and ξ ranges over particles. In English, this says: "If ξ is any particle whose mass is M, whose instantaneous acceleration is $\mathbf{A}$, and which is acted upon by a force $\mathbf{f}$, then $\mathbf{f} = M\mathbf{A}$." While logicians may find this expression of the law more revealing, we shall find the abbreviated form, $\mathbf{F} = m\mathbf{a}$, sufficient for most of our purposes. (We shall have occasion later—in Chapter 5—to use still another formulation of the law.)

---

[1] It has been argued that this is not what Newton intended under the title of "second law of motion," but I shall follow the customary practice of referring to '$\mathbf{F} = m\mathbf{a}$' as 'Newton's second law of motion'. See Brian Ellis, "The Origin and Nature of Newton's Laws of Motion," in R. Colodny (ed.), *Beyond the Edge of Certainty* (Prentice-Hall, 1965).

We have, then, a law of nature in the form of a *vector* equation relating the force, **F**, acting upon a body, the body's mass, *m*, and its instantaneous acceleration, **a**. This vector equation can be broken down into three component (or *scalar*) equations as follows:

$$\mathbf{F}_x = m\mathbf{a}_x = m\,\frac{d\mathbf{v}_x}{dt} = m\,\frac{d^2x}{dt^2}$$

with similar equations for *y* and *z*.

Translation of the vector equation $\mathbf{F} = m\mathbf{a}$ into three scalar equations is usually necessary in the solution of specific concrete problems. As we shall see, it is the components of the various vectors involved, along the three coordinate axes, which are actually measured in most applications of the law.

(If we set $\mathbf{F} = 0$ in Newton's second law, we can formally derive Newton's *first* law of motion, that is, a body upon which no forces act will move uniformly—at constant speed in a straight line.[2] Newton's *third* law of motion states that when a body, *A*, exerts a force, $\mathbf{F}_{AB}$, on a body, *B*, then body *B* will exert a force $\mathbf{F}_{BA}$ on *A* given by $\mathbf{F}_{AB} = -\mathbf{F}_{BA}$.)

We are assuming that '$\mathbf{F} = m\mathbf{a}$' is a substantive empirical statement rather than a definition of '**F**' in terms of '*m*' and '*a*'. This was clearly Newton's intention, as well as the view held by most physicists and philosophers during the period leading up to the development of the theory of rela-

---

[2] Why, then, it may be asked, did Newton state this apparent *consequence* of the second law as a separate law of motion? A full answer to this question will require a deeper discussion of the concept of a frame of reference than was given in Chapter 1, and will therefore be postponed to a later chapter. (See Chapter 12.) I shall only remark now that in concrete problem-solving situations it is sufficient to have the second and third laws of motion.

tivity. Since our eventual purpose in presenting classical dynamics is to set the stage for a presentation of relativistic dynamics, this historical fact provides a sufficient reason for proceeding in this manner. Further reasons will be adduced as we go along in this and the next chapter.

Forces, then, are to be conceived as *causes* of acceleration in bodies, rather than as a shorthand way of referring to the product of mass and acceleration. This way of regarding the concept of force is necessary (and possible) because in concrete situations there are in fact means for determining the magnitude and direction of the forces acting on a given body which are independent of any reference to $\mathbf{F} = m\mathbf{a}$. That is, we need not first measure $\mathbf{a}$ and m in order to determine $\mathbf{F}$. The force acting upon a given body will in general depend upon various properties of the body and various aspects of the situation in which the body finds itself. For the present, these factors may be roughly classified as follows: (We shall investigate them in more detail in Chapter 3.)

The nature of the body (its mass, electric charge, etc.);
The nature of other bodies in its vicinity;
The distance of the body from others in its vicinity;
The velocity and acceleration of the body with respect to others in its vicinity;
Certain universal constants and other coefficients.

These various factors allow one to express various kinds of *force laws*, that is, various laws stating what the force acting is, given the specific situation.

For example, if a body A with mass m_A is at a distance r_{AB} from body B with mass m_B, B will experience a force due to A whose magnitude is given by

$$f_{AB}^{g} = G\,\frac{m_A\,m_B}{r_{AB}^2}$$

(where G is a constant) and whose direction is in the straight line joining A and B. This is Newton's law of gravitation. If A and B are also electrically charged, having charges e_A and e_B, B will be acted upon by a force due to the presence of A whose magnitude is given by

$$f_{AB}^e = C \frac{e_A e_B}{r_{AB}^2}$$

(where C is a constant). This is Coulomb's law. The total force on A due to the presence of B (if there are no other force-determining properties in the situation) will be given by the sum of the above two expressions:

$$f_{AB}^{total} = f_{AB}^g + f_{AB}^e = \frac{Gm_A m_B + Ce_A e_B}{r_{AB}^2}$$

Another force law, known as Hooke's law, gives the force acting upon a body attached to a spring as shown in the figure below:

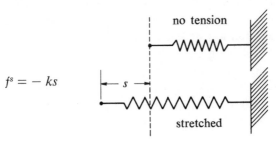

$$f^s = -ks$$

The distance of the body from the untensioned position is s, and k represents the "stiffness coefficient" for the spring. (If k is large, the spring is stiff, and the "restoring force" f^s is large; a small value for k expresses the fact that the spring is weak.) The force always acts to restore the body to the untensioned position, and its magnitude is proportional to the distance of the body from that position.

If we know the force law for a particular type of situation —how the force on the body in question depends on the factors present in that situation—we can use it, together with $\mathbf{F} = m\mathbf{a}$, to calculate how the body will move in that kind of situation. This is what one typical sort of problem in classical dynamics amounts to. Many exercises in mechanics textbooks are of this type. We may schematize this kind of problem as follows:

$$\left.\begin{array}{l} \text{Given:} \quad \mathbf{F} = m\mathbf{a} = m\,\dfrac{d^2\mathbf{r}}{dt^2} \\[2ex] \text{Given:} \quad \mathbf{F}_{\text{situation}} = \mathbf{f}\ (\text{parameters of situation}) \end{array}\right\} \begin{array}{l} \text{combine} \\ \text{to obtain} \end{array}$$

$$\mathbf{f} = m\,\frac{d^2\mathbf{r}}{dt^2} \longleftarrow$$

Problem: Solve the last equation for $\mathbf{r}\,(t)$.

We are given $\mathbf{F} = m\mathbf{a}$, the law which states what forces *generally* do: they produce or cause accelerations. We are also given an equation expressing what the *particular* force is in the situation at hand (the "force law" for the situation). Combining these two equations yields an equation relating the particular force function, $\mathbf{f}$, to the motion of the body. We must then solve (integrate) this differential equation to compute the position of the body as a function, $\mathbf{f}'$, of time (and of the parameters entering into the given force law):

$$\mathbf{r} = \mathbf{f}'\ (t,\ \textit{various parameters of the situation})$$

If we already know the motion of the body, we have *explained* its motion on the basis of $\mathbf{F} = m\mathbf{a}$ and the particular force law; if we did not already know the motion, we have *predicted* it on the same basis. In either case we have related the motion to those aspects of the situation which are responsible for it.

Problems of this sort may involve some rather difficult mathematical manipulations, depending on the complexity of the particular force function, **f**. Nevertheless, they are essentially exercises in applied mathematics, and in this sense may be said to be *physically* trivial. There are always approximative procedures that can be used if the differential equation in question is mathematically too complex to yield to a manageable explicit solution.

In order to get a better feel for this sort of *mechanical problem*, let us look at an example in which the necessary mathematical manipulations are relatively easy, the case of a body at the end of a spring.

AN EXAMPLE

We stretch the spring D units from its untensioned position and let go at time $t = 0$ (rather than, say, "giving the body a push" to get things going).

Filling in the schema we had earlier, we are given $\mathbf{F} = m\mathbf{a}$, stating what forces generally do; and we are given the force law for this kind of situation, Hooke's law,

$$f^s = -ks$$

Combining these two equations as before, we get

$$-ks = m\frac{d^2s(t)}{dt^2} \qquad \ldots (1)$$

a differential equation stating how the instantaneous acceleration of the body depends upon where it is (s). We must solve (integrate) this equation to find explicitly how s varies with time.

By mathematical ingenuity, or by consulting a handbook of solutions to various kinds of differential equations, we learn that eq. (1) has a family of solutions, given by

$$s(t) = \alpha \cos \sqrt{\frac{k}{m}}\, t \pm \beta \sin \sqrt{\frac{k}{m}}\, t \qquad \ldots (2)$$

That is, if we put in various numbers at random for α and β, a *set* of functions, $s_{\alpha\beta}\,(t)$, each of which is a solution to eq. (1), will be generated. This represents the fact that we have not yet *used* any information about how far the spring was initially stretched, and whether it was then let go or "pushed." But in the given situation the spring was stretched until, at time $t = 0$, $s = D$. Putting this information (this *initial condition*) into eq. (2), we get

$$\text{at } t = 0, s\,(0) = D = \alpha \cos \sqrt{\frac{k}{m}} \cdot 0 + \beta \sin \sqrt{\frac{k}{m}} \cdot 0$$

or

$$\text{since } \cos 0 = 1, \sin 0 = 0; D = \alpha$$

Using this information about the position of the body at $t = 0$, the family of solutions has been narrowed to

$$s(t) = D \cos \sqrt{\frac{k}{m}}\, t + \beta \sin \sqrt{\frac{k}{m}}\, t \qquad \ldots (3)$$

The β still allows for an infinite number of solutions. But of course the body does only one thing: it executes one determinate motion. We now use the knowledge that the body was simply released at $t = 0$, rather than having some initial velocity imparted to it. That is, at $t = 0$, $v(t) = ds/dt = 0$. First we differentiate eq. (3) to obtain

$$\frac{ds}{dt} = v = \sqrt{\frac{k}{m}} \left[-D \sin \sqrt{\frac{k}{m}}\, t + \beta \cos \sqrt{\frac{k}{m}}\, t \right]$$

So, at $t = 0$ we have $v(t) = 0$, and this becomes

$$0 = \sqrt{\frac{k}{m}}\,[0 + \beta]$$

or

$$\beta = 0$$

Finally, then, the equation which expresses how the body will move in the given kind of situation and subject to the given initial conditions is

$$s(t) = D \cos \sqrt{\frac{k}{m}}\, t \qquad \qquad \ldots (4)$$

s varies sinusoidally as time goes on with a period of oscillation given by

$$T = 2\pi \sqrt{\frac{m}{k}} \ \left(\text{setting} \sqrt{\frac{k}{m}}\, T = 2\pi\right)$$

A plot of s as a function of time looks like this:

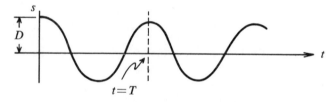

$$t = T$$

The body oscillates back and forth, from D to $-D$. (The reader should check the way in which the period $T = 2\pi\sqrt{m/k}$ varies with k and m against his intuitions, recalling that a large value of k means a stiff spring.)

Our result, expressed in eq. (4), (almost) agrees with experience. We have explained (or predicted) the motion of the body in the given situation by relating the motion to its cause, the restoring force of the spring. The formal structure of the solution to this problem conforms to what has been called the *covering law model of explanation*. We have used *general laws* (Newton's second law of motion plus Hooke's law) together with *initial conditions* to *deduce* a statement describing the motion of the particular body.[3] The 'almost' above is important. For we know from experience that the oscillations of a body on a spring usually tend to die out. That is, eq. (4) does not accurately describe the observed motion. It will be instructive to see how this discrepancy can be handled by looking at a similar but slightly more complex problem.

Suppose that the spring is immersed in a viscous fluid, or is subject to some other sort of dissipative or *damping force*, that is, a force whose magnitude is proportional to the velocity of the body and oppositely directed. We thus introduce further aspects of the situation, with their appropriate force laws. We have now, in addition to force of the spring on the body, a damping force given by

$$f^d = -cv = -c\frac{ds}{dt}$$

where c is a constant expressing the resistance of the fluid. The total force on the body is therefore

$$f^{total} = -ks - c\frac{ds}{dt}$$

Using this together with $\mathbf{F} = m a$ as before, we get

$$m\frac{d^2s}{dt^2} = -ks - c\frac{ds}{dt} \qquad \ldots (5)$$

Using the same initial conditions as before, together with a

---

[3] For a full description of the covering law model of explanation see C. G. Hempel, "Aspects of Scientific Explanation," in a book by the same title and author (The Free Press, 1965).

handbook of solutions to differential equations, we learn that the solution to eq. (5) is

$$s(t) = D\,e^{-\frac{ct}{2m}} \left\{ \cos \sqrt{\left(\frac{k}{m} - \frac{c^2}{4m^2} \right)}\; t \right\} \quad \dots (6)$$

If $c = 0$ (vanishing damping force), eq. (6) will reduce to the original solution expressed in eq. (4), as should be expected. In this sense we have in eq. (6) a solution to a *more general problem* than we had at first. The term in brackets represents an oscillation, but one whose frequency depends not only on k (the stiffness of the spring) but also on c (the viscosity of the fluid). The exponential term outside the brackets represents a *decay* of the magnitude of the

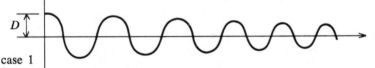

case 1
$k >> c$ Oscillation with gradual decay; '$k >> c$' means spring is very stiff compared to viscosity of medium.

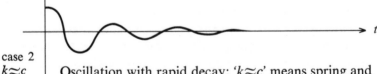

case 2
$k \approx c$ Oscillation with rapid decay; '$k \approx c$' means spring and medium about "equal in effect."

case 3
$k << c$ No oscillation, decay only. '$k << c$' means spring very weak compared to viscosity of medium. ($\sqrt{(\dots)}$) is less than 0, so that $\cos \sqrt{(\dots)}\,t$ becomes nonoscillatory.

oscillations, and the rate of decay depends on c. All of this agrees qualitatively with experience. We can go further and plot eq. (6) for various relative values of k and c, and check these quantitative results against our experience of such systems. (See page 23.) The three cases do conform with experience to a much closer degree than did the solution to the original problem, where damping forces were not taken into account.

If there were further discrepancies, we would look for further aspects of the situation and use the force laws appropriate to these new factors, constructing a new expression for the total force and using it together with $\mathbf{F} = m\mathbf{a}$, as before, to compute a more refined solution. But we have gone far enough to show how the schema described earlier works in specific cases to yield explanations (or predictions) of the motions of bodies when subject to various kinds of forces.

OTHER PROBLEMS

These have been relatively easy problems to solve. More complex (and more interesting, less "textbookish") problems arise when we consider systems containing several mutually interacting bodies. The solar system provides an example. In this system each body exerts a force on every other body, given by Newton's law of gravitation. For each pair of bodies, i and j, there will be two vector equations:

$$\mathbf{F}_{ij} = m_j \mathbf{a}_j$$

and

$$\mathbf{f}_{ij}^{\,g} = -\, G \frac{m_i m_j}{|\mathbf{r}_{ij}|^3} \mathbf{r}_{ij}$$

where the 'ij' subscript indicates the force on j due to the presence of i. We therefore have a *system* of interrelated differential equations of the form

$$m_j \frac{d^2 \mathbf{r}_j}{dt^2} = - G \frac{m_j m_i}{|\mathbf{r}_{ij}|^3} \mathbf{r}_j$$

to solve for $\mathbf{r}_j(t)$ (where $j = 1, 2, 3, \ldots$ number of bodies in the system).[4] The solution of such a system of equations is notoriously difficult to obtain. Volumes have been written presenting various approximative techniques for extracting usable information from such systems of equations. For example, there are mathematical techniques for answering certain interesting questions about the system of bodies without having a full solution for the precise motions, questions such as, Is the system as a whole stable, or will the bodies involved eventually "fly apart"? What are the periodicities involved?

Mechanics also deals with systems in which the size and shape of the bodies involved cannot be neglected, in which the diameters involved are not small in comparison with the mutual distances. If the bodies under consideration nevertheless preserve their size and shape, we have that branch of mechanics known as *rigid body dynamics*. This can be developed from *particle dynamics* by a limiting process which essentially treats a rigid body as a system of particles constrained to maintain fixed separations from each other. Thus, we can "reconstruct" the concept of a rigid body of

---

[4] The reader who is puzzled by the appearance of the $|\mathbf{r}_{ij}|^3$ term, having expected an inverse-*square* law of gravitation, deserves an explanation. The expression $\mathbf{r}/|\mathbf{r}|^3$ *does* represent an inverse square law, since the magnitude of $\mathbf{r}$ is $|\mathbf{r}|$, giving $|\mathbf{r}|/|\mathbf{r}|^3$, which is equal to $1/|\mathbf{r}|^2$. But the more complex way of expressing the law has the advantage that it reflects the *direction* of the force as well.

any given size and shape (and density distribution) in terms of the concepts of particle dynamics as follows:

Consider a system of particles (point-masses) whose relative distances are fixed.

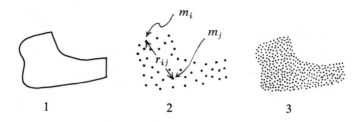

1   2   3

We imagine the particles of this system (figure 2) arranged in the same general shape as the given rigid body (figure 1), the sum of their masses being equal to the mass of the rigid body, and their distribution roughly reflecting the density distribution of the original body. These *constraints* can be expressed mathematically, with the resulting equations forming part of the system of equations to be solved for the motions of the particles with time:

$$|\mathbf{r}_{ij}| = c_{ij}, \text{ a constant; } i,j = 1,2,3, \ldots n \text{ (number of particles)}$$

$$\sum_{i=1}^{n} m_i = M \text{ (mass of original rigid body)}$$

If the result of such a reconstruction does not lead to a solution which agrees with experience as closely as we would like, we could proceed to a more accurate reconstruction (figure 3). In practice this cumbersome method is not actually used, but analogous mathematical techniques do the work of going directly to the limit of the original rigid body.

These techniques allow one to deal with systems such as a rotating top (where, obviously, size and shape are crucial), and the detailed motions of the earth-moon system (to name only two examples).

Notice that for the purposes of offering mechanical explanations of the motions of rigid bodies, the constraints involved are *assumed*. In this context no attempt is made to explain why the mutual separations of the particles are constant. Indeed, we are dealing, not with real particles, but with a sequence of idealizations in a limiting process. Nor does rigid body dynamics have anything to say about *why* rigid bodies are rigid. This is a problem for "matter theory." (Of course, matter theory may tell us that the *actual* constituents of a rigid body are entities to which, say, particle dynamics applies; or, perhaps, entities to which rigid body dynamics itself applies.[5])

One can also handle systems containing "not-quite-rigid" bodies: *elastic bodies*. These are bodies which almost preserve their shape; that is, they can undergo deformations in shape and return to normal (a golf ball, for example). The concept of such a body is reconstructed on the basis of the concepts of particle dynamics as before, except that the mutual distances, r_{ij}, of the particles are not constant. If we let 'Δr_{ij}' represent a displacement of r_{ij} from its normal value due to a force acting on i by j, then we assume as an equation of constraint

$$\mathbf{f}_{ij} = -k_{ij}\,\Delta \mathbf{r}_{ij}$$

This is just Hooke's law; r_{ij} will oscillate sinusoidally. We have, essentially, a system of particles attached to each other by springs (locally—k_{ij} is assumed to be small for large r_{ij}). Once again, there are shortcut mathematical techniques for this procedure. If the equations of constraint also in-

---

[5] This possibility seems logically odd, suggesting an infinite regress. It is like saying that Chlorine gas is green because it is composed of green atoms. But, after all, is it not an *empirical* fact that this is not the case?

cluded a damping force term, we would be reconstructing "not-quite-elastic" bodies.

The principle should now be clear. By introducing various types of *equations of constraint* the conceptual apparatus of *particle dynamics* can be used to mimic the behavior of other kinds of real mechanical systems, including *liquids* and *gases*.

Once we have carried out procedures such as those just outlined, thereby constructing the basic concepts of what is called *continuum mechanics*, these latter concepts can be taken as logically primitive, and particle mechanics can then be recovered as a limiting case of continuum mechanics. (The concept of a particle would be construed as a small region of very high mass-density, to put it roughly.) These two directions of analysis do not conflict, as they may appear to do at first glance. The move from particle mechanics to continuum mechanics is a constructive procedure, involving certain limiting processes, in which the concepts of particle mechanics are used to construct concepts of the basic entities of continuum mechanics. The entities involved in succeeding stages of the limiting process are purely fictional. The logical move from continuum mechanics to particle mechanics, however, can be conceived more realistically, for a particle can be considered to "really be" a very small rigid body.

The kinds of problems we have been discussing are largely constitutive of the work that occupied physicists throughout the eighteenth and nineteenth centuries. The elaboration of solutions to these problems forms a large part of what T. S. Kuhn has called doing "normal science" in the field of mechanics.[6] Solutions to problems such as

[6] See T. S. Kuhn, *The Structure of Scientific Revolutions* (University of Chicago Press, 1962).

these not only are mathematically complex but also involve conceptual difficulties, for the Newtonian formulation of classical mechanics is not always easy to apply. This fact led a number of physicists to attempt to reformulate the basic laws of classical mechanics in a more perspicuous form, giving rise to a different sort of mechanical problem, which Kuhn has aptly called "clarification by reformulation."[7] This highly theoretical branch of mechanics, known as "analytical mechanics," occupied men such as d'Alembert, Lagrange, Hamilton, Jacobi, Poisson, Hertz, Mach, and Poincare. Since their reformulations of mechanics did not play a crucial role in the development of relativistic mechanics, I shall not discuss them here. The Newtonian laws of motion can be recovered from the more abstract formulations, and a clear understanding of the nature and use of *these* laws is sufficient as a foundation for understanding the nature and significance of relativity theory. (For further remarks about analytical mechanics see Chapter 12.)

[7] *Ibid.*, p. 32.

3

CLASSICAL DYNAMICS
PART 2

"If everything is permissible to me . . . , if nothing
offers me any resistance, then . . . I cannot use
anything as a basis, and consequently every
undertaking becomes futile. . . . My freedom . . .
consists in my moving about within the narrow
frame that I have assigned myself. . . ."

Stravinsky

FIND-THE-MOTION PROBLEMS

We saw in the last chapter that a large part of problem
solving in classical dynamics consists in the following procedure:

Examine the system of bodies at hand in order to determine the particular kinds of forces operative, and write
down the appropriate force laws. Combine these laws with
Newton's second law of motion and with the appropriate
initial conditions (and equations of constraint, if needed),
and solve the resulting system of differential equations for
the motions of the bodies of the system: determine explicitly
$r(t)$ for each body of the system. We then have an explanation (or prediction) of the motions of the bodies under the
influence of the forces known to be acting.

We also saw that the forces acting on a body depend
upon various factors present in the situation. A classification
of the kinds of factors determinative of forces would amount

to a description of a large part of the empirical content of classical dynamics. This undertaking would be appropriate to a physics textbook on mechanics, and will not be presented here, since our purpose is to present the conceptual foundations of the theory. However, it is essential to see in outline what such a description would look like.

The various factors which have been found *by experience* to be relevant in determining the force acting upon a body i can be listed in schematic form as follows:[1]

$\mathbf{r}_{ij}$ position of i with respect to other bodies in system ($j = 1, 2, \ldots$, number of bodies in system)

$\dot{\mathbf{r}}_{ij}$ velocity of i with respect to other bodies in system

$\ddot{\mathbf{r}}_{ij}$ acceleration of i with respect to other bodies in system

m_i the mass of i

m_j the masses of other bodies in system

c_k certain universal constants

b_k certain coefficients depending on the problem at hand (including, for example, any electric charges present)

Notice that certain factors which may be present in any real system have been (quite purposefully) omitted from the list: for example, the colors of the bodies of the system, their temperatures, and their cost (which should surprise no one). These are *nonmechanical factors or parameters.* We have learned *from experience* that they have no influence on the forces acting, and are therefore to be ignored for purposes of accounting for the motions of a system of bodies.[2]

---

[1] I shall now use a notational convention which has been found useful in saving ink:
A dot placed over a symbol representing some parameter signifies the first derivative of that parameter with respect to time, two dots the second derivative, etc. Thus, $\dot{\mathbf{r}} = dr/dt$, $\ddot{\mathbf{r}} = d^2r/dt^2$, etc.

[2] In the cases of some nonmechanical parameters the knowledge that they are not determinative of forces was historically hard won. It is only by hindsight that we might smile at the possible inclusion

With the aid of the above list of *mechanical parameters* (force determinative factors) we can write a general equation or schema for the force law for *any* situation:

$$\mathbf{f}^{\text{total}} = \mathbf{f}\,(\mathbf{r}_{ij}, \dot{\mathbf{r}}_{ij}, \ddot{\mathbf{r}}_{ij}, m_i, m_j, c_k, b_k)$$

In English, this says that the total force acting on a body *i* is equal to some function of the parameters within the parentheses. In given particular situations, of course, the force function will be much simpler than the general schema may suggest. Let us list some examples.

Newton's law of gravitation (for a two-body system):

$$\mathbf{f}_m^g = -\,G\,\frac{mM}{|\mathbf{r}_{mM}|^3}\,\mathbf{r}_{mM} = \mathbf{f}\,(\mathbf{r}_{mM}, G, m, M)$$

$$\underset{\mathbf{r}_{ij}}{\uparrow}\;\underset{c_k}{\uparrow}\;\underset{m_i}{\uparrow}\;\underset{m_j}{\uparrow}$$

Galileo's "law of gravity" (near the surface of the earth):

$$\mathbf{f}_m^G = mg = \mathbf{f}\,(m,\,g)$$

$$\underset{m_i}{\uparrow}\;\underset{b_k}{\uparrow}$$

Hooke's law:

$$\mathbf{f}_m^s = -\,k\mathbf{s} = \mathbf{f}\,(k,\,\mathbf{s})$$

$$\underset{b_k}{\uparrow}\;\underset{\mathbf{r}_{ij}}{\uparrow}$$

The "law of damping":

$$\mathbf{f}_m^d = -\,c\mathbf{v} = \mathbf{f}\,(c,\,\mathbf{v})$$

$$\underset{b_k}{\uparrow}\;\underset{\mathbf{r}_{ij}}{}$$

---

of them in the list of mechanical parameters. It is not *a priori* obvious, for example, that the color of a body is irrelevant to the force it may exert on other bodies.

Coulomb's law (for two charged bodies):

$$\mathbf{f}_m^e = C\,\frac{e_m\,e_M}{\mathbf{r}_{mM}^2} = \mathbf{f}\,(\underset{\underset{b_k}{\nwarrow\;\nearrow}}{e_m,\,e_M},\,\underset{\underset{c_k}{\uparrow}}{C},\,\underset{\underset{\mathbf{r}_{ij}}{\uparrow}}{\mathbf{r}_{mM}})$$

A more complex force law, with which we shall later be concerned, is the Lorentz force law:

$$\mathbf{f}^{\mathbf{EH}} = e_m\left[\mathbf{E} + \frac{1}{c}\,(\dot{\mathbf{r}}\times\mathbf{H})\right] = \mathbf{f}(\underset{\underset{\dot{\mathbf{r}}_{ij}}{\uparrow}}{\dot{\mathbf{r}}},\,\underset{\underset{c_k}{\uparrow}}{c},\,e_m,\,\underset{\underset{b_k}{\nwarrow\;\uparrow\;\nearrow}}{\mathbf{E},\,\mathbf{H}})$$

This expresses the resultant force on a charged body in a region of electric and magnetic fields in vacuum. (We shall see in Chapter 4 that this basic law of *electrodynamics* is, from the point of view of classical dynamics, just one more force law.)

Complex situations will of course call for various combinations of the above force laws to determine the motions of the bodies of the system. We had such an example in the last chapter: the mass at the end of a spring, subject also to the resistance of the surrounding medium.

Mechanics textbooks abound with problems in which the situation is described, and the student is to apply the appropriate combination of force laws, together with $\mathbf{F} = m\mathbf{a}$ and the initial conditions, to find $\mathbf{r}(t)$ for each body of the system. A "handbook of basic situations" which would list the force laws appropriate to various basic kinds of situations could be constructed, and one could simply look up the appropriate laws for any given problem.

FIND-THE-FORCE-LAW PROBLEMS

There is another rather different and very basic sort of problem in mechanics that is related to the one we have just discussed as its inverse. Suppose that we are *given the motions* of the bodies of a system, as well as any parameters of the system we may ask about. The problem is: *Determine the force laws* that are operative and responsible for the given motions. That is, we are given $r(t)$ for each body (and of course $F = ma$), and asked to find $f(various parameters)$. We could describe this as the "draw-up-the-handbook" problem. This type of problem is characteristic of certain types of basic research in physics in general, not only classical dynamics. Thus, for example, in nuclear physics, where Newton's laws of motion (including $F = ma$) do not even apply, one still speaks of the problem of determining the "nuclear forces" which account for the motions —or better, "interactions"—of nucleons.[3]

With reference to classical dynamics, an example of this type of problem is Newton's attempt to formulate the force law responsible for the observed motions of the planets. These are "why does the system behave as it does?" problems, and answers are of the form: "because the force law for systems of this type are. . . ." We verify a purported answer to this kind of problem by solving the inverse problem discussed earlier: we use the purported answer, together

---

[3] The problem-solving schema of Chapter 2 can be nicely generalized to areas of physics other than classical mechanics. Instead of talking about the general force law, $F = ma$, we introduce the notion of a general law expressing the temporal development of the *state* of a system (the Schrodinger equation in elementary quantum mechanics, for example). And instead of speaking of particular force laws for particular situations, we introduce the notion of appropriate *state-functions* for particular sorts of systems. Finally, instead of solving for the motions of bodies, we solve for the evolution with time of certain well-defined parameters (which may include motion in some cases).

The possibility of this generalization seems to account, in part, for the naturalness in referring to certain aspects of microphysics as 'quantum *mechanics*'. More will be said about this in Chapter 12.

with $\mathbf{F} = m\mathbf{a}$, to see if one can in fact deduce the given motions.

Now it turns out that the find-the-force problem, as I have stated it *so far*, admits of trivial solutions! For if we are given the positions of the bodies as functions of time, and $\mathbf{F} = m\mathbf{a}$, it is always possible in a trivial way to calculate back to *some* function of the parameters of the system which will both satisfy $\mathbf{F} = m\mathbf{a}$ and yield the given motions. For one body, given $\mathbf{F} = m\mathbf{a} = m\ddot{\mathbf{r}}$ and the observed motion $\mathbf{r}(t)$, we can always construct a function of $\mathbf{r}$ and the set P of the parameters of the situation, $\mathbf{f}(\mathbf{r}, P)$, such that $\mathbf{f}(\mathbf{r}, P) = m\ddot{\mathbf{r}}$. (For example, $m\ddot{\mathbf{r}}$ itself will do the trick!) But not all such functions are considered to be solutions to the original mechanical problem, although they may well satisfy an engineer, whose purpose may be simply to have a "formula" for calculating the motion when plugged into the left side of $\mathbf{F} = m\ddot{\mathbf{r}}$.

So, the find-the-force problem becomes physically interesting—not a mere mathematical exercise—only when certain *limitations* or "constraints"[4] are placed upon the allowable kinds of force functions and parameters which may appear in these functions. We find here an (implicit) element of *convention* as to what constitutes a *mechanical explanation* of the behavior of a system of bodies (rather than an engineering formula). One attempts to *reduce* some (mathematically satisfactory) force laws to others which are simpler, more basic, more intelligible, more "purely mechanical"—that is, more in accord with our assumptions, distilled from past experience, regarding the basic kinds of forces operative in the universe. An investigation of the kinds of constraints conventionally imposed by the physicist would result in a delineation of a further aspect of what

---

[4] This use of the term 'constraint' differs from its use in expressions such as 'equation of constraint', as will presently become clear.

Kuhn[5] would call the classical mechanics "paradigm"—that set of fundamental (and perhaps tacit) assumptions which guides research and tells us when we have a *mechanical* problem or its solution before us. (It should be clear that what I have called "conventionally" imposed constraints are not necessarily arbitrary or unreasoned, for they are in large measure distilled from past usage.) Let us clarify some of these ideas by way of an example.

Consider once again the mass attached to a spring, and let us suppose that the force law

$$f = -ks - cv$$

did not, in conjunction with $\mathbf{F} = m\mathbf{a}$, accurately reproduce the observed motion. We would then look for other aspects of the situation and their force laws. But we could quite easily generate the observed motion to any desired degree of accuracy by constructing a "force function" as follows:

Write down the series

$$f = -ks + a_2s^2 + a_3s^3 + \ldots + a_ms^m + \ldots$$
$$-c\dot{s} + b_2\dot{s}^2 + b_3\dot{s}^3 + \ldots + b_m\dot{s}^m + \ldots$$

We now simply adjust the coefficients a_i and b_i in such a way as to give a "solution" (a predicted motion $s(t)$) which agrees as closely as we wish with the observed motion. If required, we could introduce a new series in $\ddot{s}$ with adjustable coefficients, or inverse powers of s and $\ddot{s}$, or *whatever seemed to be required*. (Most of the coefficients would in fact be $= 0$, so the final "solution"—the "force function" f—would not be as messy as the general formula for the series might suggest.) This *function-fitting* procedure obviously does not yield a mechanical *explanation* of why

[5] T. S. Kuhn, *The Structure of Scientific Revolutions*.

the mass-spring system behaves as it does.[6] Why not? Because there is no way to *physically interpret* all of the nonzero coefficients in the series. They cannot be related to specific aspects of the actual system as expressing the *sources*, in that situation, of the forces acting on the mass. The coefficients k and c (which are the a_1 and b_1 terms) *can* be so interpreted, as we saw in the previous chapter; k represents the stiffness of the spring and c represents the viscosity of the surrounding fluid. As such we can trace these two terms of the force function directly to properties or parts of the system, but not the other terms. (We need not even have recourse to the device of a series with adjustable coefficients to show that not all mathematically satisfactory force functions yield mechanical *explanations*. See Appendix A for a different demonstration of this fact.)

At this point it might be argued that my claim that the parameters of a *physically* satisfactory force law must be physically interpretable—traceable to identifiable aspects of the situation—is unconvincing because, say, it smacks of a certain kind of metaphysics that we have learned to expunge from science. While I do not believe that this generally "positivistic" attitude accurately reflects the nature of problem solving in mechanics, I think it can still be shown, apart from my claim about the interpretability of the parameters, that some limitations must be imposed on the allowable set of force functions.

For if we ignore (or deny) the fact that k and c are interpretable as explained above, then why should $f = -ks-c\dot{s}$ be considered as yielding a mechanical explanation (if it had been mathematically satisfactory), or $f = -ks$ (had it been otherwise correct)? Or, why should Newton's law of gravitation be considered as giving an explanation of the

---

[6] Once again, the engineer's purpose *would* be fulfilled by this procedure.

motions of the planets? Generally, why should *any* accepted force law be considered as yielding a *mechanical explanation* rather than only a mathematically correct *formula* or mere prediction scheme which an engineer would find satisfactory? Why not always simply use a power series with adjustable coefficients?

Why indeed! Some philosophers (and physicists) would simply say that in principle *any* mathematically adequate "force function" is acceptable—that explanation in mechanics *is* only the finding of a function that gives back the motion when plugged into the left side of $\mathbf{F} = m\mathbf{a}$. If some mathematically adequate functions are frowned upon, this is merely the physicist's habit or prejudice in preferring to use *simple, familiar* functions which have worked in the past.

There is a sense in which this positivistic view cannot be "refuted," for there is no *a priori* reason why certain simple and familiar kinds of force laws should be considered more physically satisfactory or explanatory than others.[7] But this is like saying that there is no *a priori* reason why the present rules of chess (say) are more satisfactory than others we might invent. Indeed, if the only way we could effect mate in a given situation were to change the rule for how the knight moves, there would be no *a priori* reasons against so doing. But this would clearly no longer be playing *chess*. Similarly, allowing any force function which would fill the bill *mathematically* in a given situation would no longer be doing *mechanics*—solving the *mechanical* problem. One could say, then, that the positivist is proposing that we do mechanics in a new way; he is not adequately analyzing mechanics as we have it.

It is past success with a limited number of simple force

---

[7] It should be kept in mind that I am waiving my claim about interpretability of parameters at this point.

functions (with physically interpretable parameters, I would add) which determines that in future problems we attempt to make do with combinations of these functions (and parameters). If a case could be made for greater systematic unity and predictive power by the use of new, unfamiliar force functions, this would be a good argument for considering these new functions as physically explanatory—not merely mathematically satisfactory.[8] (Although we would still feel uneasy if the parameters of the function did not meet the interpretability condition.) We shall see in the next chapter that Maxwell's equations and the Lorentz force law provide an example of this.

If in a given kind of situation a previously unfamiliar and (thus) intuitively *ad hoc* force function predicts correctly the motions of the bodies of the system, and repeated attempts to reduce this force function (and its associated parameters) to those which are more familiar meet with failure, the new function may come to be viewed as basic.

A simple and clear example of how one force law and its associated parameters may be reduced to another is the case in which Galileo's "law of gravity," $f^G = mg$, is reduced to Newton's law of gravitation. We have

$$f^g = G \frac{mM}{R^2}$$

as the law of gravitation for a body of mass m near the surface of the earth, where M is the mass of the earth and R the radius of the earth. We assume that the distance of m from the center of the earth varies only slightly from R—

---

[8] Following the chess analogy: It is past success (in terms of interesting games) that determines that we make do with the present rules of chess. If a case could be made for more interesting games with new rules, this would be a good argument—the only good argument—for changing to the new rules.

that m does not fall from a great height. The quantity GM/R^2 therefore is a constant, which turns out to be numerically equal to g, so we have $f^g = mg$, Galileo's law. At the same time one coefficient, g, has been reduced to others: G, M, and R.

In the last "Question" of his *Opticks*[9] Newton stated the *reductive program* as follows:

"Have not the small Particles of Bodies certain Powers, Virtues, or Forces, by which they act at a distance . . . for producing a great Part of the Phaenomena of Nature? For it's well known, that Bodies act upon one another by the Attractions of Gravity, Magnetism, and Electricity; and these Instances show the Tenor and Course of Nature, and make it not impossible but that there may be more attractive powers than these." Perhaps even these are not the most basic, for "What I call Attraction may be performed by Impulse, or by some other means unknown to me." Finally, ". . . to derive two or three general Principles of Motion [force laws] from Phaenomena, and afterwards to tell us how the Properties and Actions of all corporeal Things follow from those manifest Principles, would be a great step in Philosophy. . . ."

The reductive program also finds its way into the writings of philosophers who were impressed by the great success of Newtonian mechanics. Hume, in his first *Inquiry*, said:

"Elasticity [encapsulated in Hooke's law], gravity, cohesion of parts, communication of motion by impulse; these are probably the ultimate causes and principles which we ever discover in nature; and we may esteem ourselves sufficiently happy, if, by accurate inquiry and reasoning, we can trace up the particular phenomena to, or near to, these general principles."

[9] See pp. 375–6 and 401–2 of the Dover edition (1952) of Newton's *Opticks*.

I have presented, then, two kinds of limitations or constraints on the set of allowable force functions: the *parameter-interpretability* condition and the *simplicity and familiarity* condition. But I do not want to give the impression that these form a precise set of necessary and sufficient conditions for distinguishing in every case in a sharp manner between *explanatory* functions and nonexplanatory functions. There are indeed cases where the two conditions conflict with each other. I am claiming only that these two restricting conditions or constraints are in fact (tacitly) at work in mechanics, and were we to ignore them completely, a very basic kind of research problem in mechanics would be trivialized.

I began this chapter with a quotation from Stravinsky. A quotation from Kant (also on a different subject) is just as apt: "The light dove, cleaving the air in her free flight, and feeling its resistance, might imagine that its flight would be still easier in empty space."

The Newtonian or mechanical view of the universe was that of particles located at various positions in space, capable of moving about, and a *small number of basic kinds of forces* acting among them. Rigid bodies, elastic bodies, and fluids are then analyzed through a mathematical device (the integral calculus) as, roughly speaking, an infinite number of particles related to each other in accordance with certain equations of constraint. Eventually, such bodies (actual bulk matter) would be treated more realistically when we come to know the actual forces acting among the (finite number of) actual "atoms" of which they are composed.[10] This came to be the task of chemistry, and even today many of the problems of nuclear physics are not too far removed

---

[10] As opposed to the fictional particles we had before going to the limit.

from being particular problems of the general sort described in this chapter.

This completes our preliminary discussion of *classical mechanics.* There are a number of matters of importance which I have not yet raised, but they are best left to later chapters when the foundations of *relativistic mechanics* will be presented.

Our next task is to introduce the fundamentals of *classical electrodynamics,* for, as we shall see, it is against the background of certain difficulties in the relation between classical mechanics and classical electrodynamics that the necessity of a replacement for classical mechanics arose.

4

ELECTRODYNAMICS

*"Wherever possible, substitute constructions out of
known entities for inferences to unknown entities."*
Russell

Newton spoke of "the Attractions of Gravity, Magnetism,
and Electricity" in the speculative last "Question" of his
Opticks, and he provided a successful theory of the first of
these "Attractions" in his *Principia.* That is, he produced a
force law—the inverse square law of gravitation—which,
when combined with his second law of motion, allows the
deduction of the motions of a group of bodies under the
influence of gravitational forces. This was perhaps the great-
est achievement of Newtonian mechanics. The only work in
this area left for physicists in the eighteenth and nineteenth
centuries was the detailed application of Newton's laws to
the complex motions of the bodies of the solar system, and
the formulation of a more sophisticated mathematical presentation of the theory.[1]

But Newton had no similar success with the "Attractions"
of electricity and magnetism. It was not until the nineteenth
century that physicists succeeded in isolating those parameters describing bodies and their environments which could
be used to construct a force function to account for the
observed motions of bodies in situations considered to be

[1] These two kinds of research were briefly discussed toward the
end of Chapter 2.

electrical and/or magnetic in character. This "delay" was in large measure due to the relative complexity of nature in this regard.

THE LAWS OF ELECTRODYNAMICS

Before presenting the relevant force law, I will first discuss the parameters that enter into it. I shall not attempt to trace the history of this endeavor, but will simply present the results.[2] We shall see that there is an important sense in which the "theory of electricity and magnetism"—or *electrodynamics*—is just one more branch of classical dynamics, in the same way in which gravitational theory, or "spring theory," are branches of classical dynamics. That is (as Newton suspected), electrodynamics, as developed in the nineteenth century, is at bottom part of the Newtonian program.

According to the theory of electrodynamics all of space, including "empty space," is pervaded by two fields, the electric field and the magnetic field, each of which may vary with time. These are *vector* fields,

[2] This is in keeping with the manner in which I presented classical mechanics in the previous chapters. There, too, I did not trace the "process of discovery" with respect to Newton's laws of motion or with respect to various specific force laws. As interesting as such a study might be, it is unnecessary for an understanding of the conceptual foundations of *relativistic* mechanics, which is, after all, our eventual goal. For this purpose it is sufficient to simply *state* classical mechanics and how it is *used* (as clearly as possible). When we finally reach the stage at which relativistic mechanics can be introduced, the presentation will be different. At that point I will show how one can *arrive at* its fundamental laws and assumptions, rather than simply stating them. (I believe this to be a revealing way of presenting the theory of relativity, since, as we shall see, this theory did arise out of certain difficulties in classical mechanics and electrodynamics.)

$$E(\mathbf{r}, t) \text{ and } H(\mathbf{r}, t)$$

That is, each point in space at each instant of time is characterized by six numbers E_x, E_y, E_z; H_x, H_y, H_z, the components of the two vectors along the axes of the chosen coordinate system.[3] (We may picture these two vector fields as a pair of arrows drawn at every point of space, where the arrows may change in direction or length as time passes.) These numbers seem to represent a "condition of space," or of some fine medium pervading space.[4]

We must also introduce the notion of *electric charge density*, which will be symbolized by $\rho(\mathbf{r}, t)$. This is a *scalar* parameter, not a vector; that is, each point in space at each instant of time is characterized by a seventh number.

The two vector fields (or the one "electromagnetic field") and the one scalar field are not independent of each other but are related in accordance with certain laws expressed by *Maxwell's equations*. These are, for empty space,

$$\nabla \cdot \mathbf{E} = \rho \qquad \qquad \ldots (1)$$

$$\nabla \cdot \mathbf{H} = 0 \qquad \qquad \ldots (2)$$

$$\nabla \times \mathbf{E} = -\frac{1}{c}\frac{\partial \mathbf{H}}{\partial t} \qquad \qquad \ldots (3)$$

$$\nabla \times \mathbf{H} = \frac{1}{c}\left(\frac{\partial \mathbf{E}}{\partial t} + \rho \mathbf{v}(\rho)\right) \qquad \ldots (4)$$

For our purposes it is not necessary to spend any time

---

[3] Different coordinate systems will of course yield different sets of numbers, different "codes." The possibility of field-free regions is also allowed, since, after all, these numbers can be zero for certain positions and times.

[4] This is important, as we shall see later in this chapter, for it is an application of what I called the "parameter interpretability condition" in Chapter 3.

explaining the meaning of these equations.[5] It is only necessary to notice that we have two vector parameters and one
scalar parameter, characterizing each point in space at every
instant of time, which can vary as wildly as one may wish
to imagine—except for the fact that they will always be
related to each other in the manner expressed by Maxwell's
equations. Methodologically these are *equations of constraint*, of the same general type discussed in Chapter 2.[6]

We are now in a position to state the electric and magnetic force laws. If a small electrically charged body whose
total charge is e[7] is in a region pervaded by electric and
magnetic fields **E** and **H**, the body will experience forces
due to these fields given by

$$\mathbf{f}^E = e\mathbf{E}$$

and

$$\mathbf{f}^H = \frac{e}{c}\left(\mathbf{v}(e) \times \mathbf{H}\right)$$

These are the electric and magnetic "Attractions" about
which Newton speculated but did not succeed in determin-

---

[5] Those who are interested may consult Appendix B. Suffice it to
say here that c is an empirical constant depending in a specified way
on the electrical and magnetic properties of the medium under consideration.

[6] There are of course differences between Maxwell's equations and
the equations of constraint discussed in Chapter 2, but they resemble
each other in this important respect: they express the interdependence of the parameters which enter into various force laws.

[7] Where the total charge e is the integral of the charge density over
the body.

The point of the word 'small' is that I am really presenting *particle electrodynamics*. I have given Maxwell's equations in their so-
called "differential form." A more realistic presentation would use
their "integral form." (See Appendix B.)

ing. They are usually expressed as one force law, known as the *Lorentz force law*:

$$f^{EH} = e \left(\mathbf{E} + \frac{1}{c} \mathbf{v} \times \mathbf{H} \right) \qquad \dots (5)$$

The point is this: the electromagnetic field is the source of a force exerted upon charged bodies.

The Lorentz force law is methodologically on a par with the other force laws we have discussed, such as Newton's law of gravitation and Hooke's law. It expresses a function of various aspects of the nature and environment of a body which, when put into the left side of $\mathbf{F} = m\mathbf{a}$, will (when integrated) give the motion of the body in the environment in question, in this case an electromagnetic environment. Once again we use Newton's second law, which tells us what effect forces have on massive bodies: they cause them to accelerate; and the Lorentz force law tells us what the specific force is in electromagnetic situations.

This point is important enough to merit a few more words. If we compare the Lorentz force law with (say) Hooke's law plus damping, $f = -kx - cv$, we will notice two differences, each of which is methodologically of no relevance. First, the Lorentz force law is more complex, in the sense that it contains two vector fields as parameters. Second, the Lorentz force law carries with it an associated set of equations of constraint, Maxwell's equations, stating the interdependence of these parameters. There is no such set of equations associated with $f = -kx - cv$ stating any interdependence of k and c. (The stiffness of the spring and the viscosity of the medium are independent of each other.) These two facts allow for a degree of richness of mathematical and empirical content in electrodynamics that is not present in "elementary spring theory," but this should not blind one from the fact that *the theory of electrodynamics is but a branch of classical dynamics.*[8]

There is *another* quite distinct sense, however, in which one might reasonably suspect that the Lorentz force law and its associated parameters are *not* on a par with other force laws and their associated parameters, the sense in which we might expect the reductive program to work with the Lorentz force law and its parameters. By way of preparing the ground for a discussion of this I will first do two things: (1) show how the Lorentz force law can serve as a basis for the reduction of another force law, Coulomb's law; (2) consider some specific important consequences of Maxwell's equations themselves.

REDUCTION OF COULOMB'S LAW

In Chapter 3 Coulomb's law of electrostatic attraction was given as an example of a force law. This law is but a special case of the Lorentz force law, in the same way in which Galileo's law of gravity is a special case of New-

---

[8] We may conclude only that electrodynamics is a more interesting branch of mechanics than elementary spring theory.

The full statement of that branch of mechanics known as the mechanics of elastic bodies, which may be considered a kind of generalization of elementary spring theory, *does* contain equations of constraint, and is of the same relative richness as electrodynamics. And it is a paradigm case of a "branch of classical mechanics."

I have perhaps belabored this point, but in my view most physics texts are far from clear in this regard. It is easy to get the impression from many of them that classical mechanics and classical electrodynamics are two quite distinct theories. The proper view, in my opinion, is that the latter is a very rich and important *part* of the former.

ton's law of gravitation.[9] Let us see how the reduction is accomplished.

Suppose that we have a spherical, uniformly charged body of total charge e located at the origin of our frame of reference. Its radius is R. That is,

if $r > R$, $\rho(r) = 0$

if $r < R$, $\rho(r) = $ a constant such that $\iiint \rho(r)\, dV = e$

We now use the first Maxwell equation to determine the intensity (magnitude) of the electric field surrounding the charge.[10] Spherical coordinates (see Chapter 1) are most convenient in this problem. Eq. (1) becomes

$$\nabla \cdot \mathbf{E} = \frac{1}{r^2} \frac{\partial}{\partial r} (r^2\, E_r) + \frac{1}{r \sin \theta}$$

$$\left[\frac{\partial}{\partial \theta} (\sin \theta \cdot E_\theta + \frac{\partial E_\phi}{\partial \phi} \cdot E_\phi) \right] = \rho(r)$$

We have a spherically symmetrical situation, so derivatives with respect to the angles θ and ϕ will be zero. Thus the second term is zero, and we get

$$\frac{1}{r^2} \frac{\partial}{\partial r} (r^2\, E_r) = \rho(r)$$

[9] There is this difference between the two cases: In a literal sense Galileo's law is false; we deduce an approximation of it from Newton's law of gravitation. Coulomb's law, however, is strictly deducible from the Lorentz force law.

[10] The first Maxwell equation is an equation of constraint relating $\mathbf{E}$ and ρ. Since $\rho(r)$ has been given, we use the equation to determine $\mathbf{E}\ (r)$. (The other Maxwell equations are not needed in this electro*static* case, since the derivatives with respect to time contained therein all become zero; that is, there will be no changing magnetic fields to contend with.) Notice that we can consider the charged body to be responsible for the field.

What is the electric field E_r at a point $r > R$, where $\rho = 0$? We get

$$\frac{1}{r^2} \frac{\partial}{\partial r} (r^2 E_r) = 0$$

or

$$\frac{\partial E_r}{\partial r} + \frac{2}{r} E_r = 0$$

The solution to this equation is

$$E_r = k/r^2$$

where k is a constant proportional to e, and so can be written as Ce, where C is another constant.

If then, we have another charged body (at rest) at r, with total charge e', the Lorentz force law states that the force acting upon it due to the field is

$$f_r = e'E_r = C \frac{ee'}{r^2}$$

which is Coulomb's law of electrostatic "attraction." "Coulomb forces," then, are not *sui generis*—just as "Galileo forces" are not. They both are the result of "more basic" forces—reducible to them.

ELECTROMAGNETIC WAVES

There is a mathematical consequence of Maxwell's equations which is of crucial importance. It is possible to derive the following equation (the "wave equation"):

$$\nabla^2 \mathbf{E} = \frac{1}{c^2} \frac{\partial^2 \mathbf{E}}{\partial t^2}$$

There is a similar equation for **H**. These are vector equations, each one of which encapsulates three scalar equations, one for each component of the vectors involved. In the rectangular coordinate system we have been using we have

$$\frac{\partial^2 E_x}{\partial x^2} - \frac{1}{c^2} \frac{\partial^2 E_x}{\partial t^2} = 0 \qquad \dots (6)$$

with similar equations for E_y and E_z. Eq. (6) states the manner in which changes of E_x over space are related to changes of E_x with time. This equation can be solved for E_x, giving a family of solutions of the form

$$E_x = e^{i(\kappa x - \omega t)} \qquad \dots (7)$$

which can also be written as

$$E_x = \cos(\kappa x - \omega t) - i \sin(\kappa x - \omega t) \qquad \dots (8)$$

where κ and ω are constants. That is, each pair of values assigned to κ and ω will yield a solution to eq. (8). But there is one restriction: it can be shown that eq. (8) yields a solution to eq. (6) only if κ and ω are related to each other by

$$\omega/\kappa = c \qquad \dots (9)$$

If we now assume that E_x can only assume real values, we can drop the second term of eq. (8), thereby obtaining

$$E_x = \cos(\kappa x - \omega t) \qquad \dots (8')$$

Let us graph this equation to obtain an intuitive grasp of what it says.

First, pick any point in space, say, along the z axis (so that $x = 0$), and see how E_x varies with time at that point. If $x = 0$, eq. (8') becomes

$$E_x = \cos(0 - \omega t) = \cos(-\omega t) = \cos \omega t$$

The graph of this is

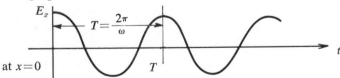

E_x varies sinusoidally with time at any given position in space, the period of oscillation being $T = 2\pi/\omega$. We can also pick some fixed time, say $t = 0$, and see how E_x varies with position at that time. If $t = 0$, eq. (8') becomes

$$E_x = \cos (\kappa x - 0) = \cos \kappa x$$

The graph of this is

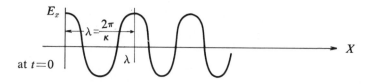

E_x varies sinusoidally with position at any given time, the length of the waveform being $\lambda = 2\pi/\kappa$. The velocity of the waveform is

$$\frac{\lambda}{T} = \frac{2\pi/\kappa}{2\pi/\omega} = \frac{\omega}{\kappa} = c$$

We see then that Maxwell's equations—the equations of constraint stating how the parameters occurring in the Lorentz force law must be related to each other—allow for the possibility of *moving electromagnetic waves* in "empty" space, with velocity c.

Further derivations from Maxwell's equations show that an oscillating electric charge will actually produce such waves, that these waves will reflect or refract when they impinge upon various kinds of material media, and that these waves will be scattered by other charges. In fact, a portion of the spectrum of electromagnetic waves behaves in exactly the same way as previously known waves of light, and they move with the same velocity. That is, the *measured* velocity of light waves in vacuum is identical with the *calculated* value of c (which depends upon the electrical and magnetic properties of various kinds of media, in this case empty space). There seems, in fact, to be no good reason for denying that light waves simply *are* (one portion of the spectrum of) electromagnetic waves. This empirical identity was in fact claimed toward the end of the nineteenth century.[11]

Now, if light consists of waves, a rather natural question arises: waves of *what*? The physical concept of a wave, after all, is that of a periodic change of some parameter referring to some property or aspect of nature. It had been speculated, even before the advent of electrodynamics, that light waves were really vibrations in some fine pervasive *elastic medium* called the "luminiferous ether." There were similar speculations regarding electromagnetic waves, which yielded the concept of an "electromagnetic ether." (E, H, ρ, and c would then refer to certain properties of, or processes in, this ether.) So, the physical identification of light waves with electromagnetic waves yields an identification of the (specu-

---

[11] This is an empirical identification of what *could* have been two distinct constituents of nature. In other words, it was *learned empirically* that the fundamental referent of the science of optics is identical with one type of entity studied in the science of electrodynamics. (The history of this identification, as well as of other matters discussed in this chapter, can be found in E. Whittaker, *A History of Theories of Aether and Electricity* [Harper Torchbook edition, 1960].)

There are philosophers who have certain scruples which I do not share regarding theory reduction, the role of models, and the nature of "theoretical terms" in general. These philosophers may find some of my remarks in this chapter to be insensitive to what they take to be important philosophic problems. My reasons for proceeding in what may appear to them to be a naively "realistic" manner can be found in my papers "Models and Theories," *British Journal for the Philosophy of Science*, August, 1965, pp. 121–142; and "Theory and Observation," *op. cit.*, May, 1966, pp. 1–20, and August, 1966, pp. 89–104.

lated) luminiferous ether with the (speculated) electromagnetic ether.

THE REDUCTION OF ELECTRODYNAMICS

The concept of elasticity is a mechanical one predating
the development of electrodynamics. It is a concept encapsulated in a generalization of Hooke's law to three-dimen-
sional material continua. (The reader will recall from
Chapter 2 that when a body is subject to a force described
by Hooke's law, a force proportional to the displacement of
the body from its "normal" position, it will execute sinusoidal vibrations about this normal position. I also indicated briefly that the branch of mechanics known as the
mechanics of elastic bodies is a generalization of "elementary
spring theory.")

Now two of the parameters of the Lorentz force law, **E**
and **H**, can undergo (interdependent) sinusoidal vibrations, as we have seen. This suggests the following program:
Show that these electromagnetic vibrations (electromagnetic
waves) are the vibrations of an underlying mechanical medium—that the electromagnetic (= luminiferous) ether is
at bottom a mechanical *elastic medium*, or ether. If this can
be accomplished, it will have been shown that the Lorentz
force law (with its associated parameters) is *not* on a par
with other force laws (and their parameters), but can be
reduced to them. We will have *reduced electrodynamics to
the remainder of mechanics*: $F = m\mathbf{a}$ plus previously
known force laws and parameters.

The parameters ρ, c, **E**, and **H** would be reduced to, constructed out of, more familiar mechanical parameters, such
as the stiffness coefficient k in Hooke's law (or an appropriate generalization of it), just as the parameter g in Galileo's law was reduced to the parameter G in Newton's law

of gravitation (plus the radius and mass of the earth, R and M). ρ, c, **E**, and **H** would thus be shown to represent, not *independent* (electromagnetic) features of nature, but rather certain complexes of the old familiar features in a postulated elastic medium. The electromagnetic ether would be conceived, roughly, as some sort of elastic jelly, where **E** and **H** would represent internal strains (for example) which could propagate with a definite velocity. Electric charge could perhaps be some sort of vortex motion of the small parts of this jelly, "knots in the ether." Disturbance of the vortex (oscillation of an electric charge) would produce elastic vibrations in the medium (electromagnetic waves). **E**, **H**, and ρ would not require a separate place in our physical ontology. The Lorentz force law would not represent a new and distinct law of nature.

We would have succeeded in carrying out Russell's maxim (in an area for which he did not intend it):

"Wherever possible, substitute constructions out of known entities for inferences to unknown entities."[12]

The "known entities" are the recognized forces and parameters of preelectrodynamics mechanics. The "unknown entities" are the supposed refere*nts* of the expressions '**E**', '**H**', and 'ρ', where these are assumed to be newly discovered aspects of nature, independent of those constituents of the universe previously admitted into physics. The "constructions" are mathematical functions of the old mechanical

---

[12] See Russell's "Logical Atomism," reprinted in R. C. Marsh (ed.), *Logic and Knowledge* (Allen & Unwin, 1956), p. 326. It has not been recognized that Russell's "supreme maxim in scientific philosophising," intended for use in certain logical-mathematical and epistemological-metaphysical contexts, provides a useful tool for understanding certain kinds of reduction in science. (I suspect that it is the kind of scruples mentioned in the last footnote that has prevented this recognition, but that is another story.)

parameters which behave in a way which satisfies the new (electrodynamic) equations, so that these new equations, including the Lorentz force law, can be deduced from the old equations: the other specific force laws of mechanics, together with their associated equations of constraint.

In other words, Russell's maxim tells us that instead of taking, for example,

$$\nabla \times \mathbf{E} = -\frac{1}{c}\frac{\partial \mathbf{H}}{\partial t}$$

and

$$\mathbf{f}^{\mathbf{EH}} = e\left(\mathbf{E} + \frac{1}{c}\mathbf{v} \times \mathbf{H}\right)$$

as new equations to be put alongside of familiar mechanical laws, we should try to construct functions $\mathbf{M}_1$, M_2, $\mathbf{M}_3$, and M_4 of already recognized mechanical parameters such that the equations

$$\nabla \times \mathbf{M}_1 = -\frac{1}{M_2}\frac{\partial \mathbf{M}_3}{\partial t}$$

and

$$\mathbf{f}^{\mathbf{M}_1\mathbf{M}_3} = M_4\left(\mathbf{M}_1 + \frac{1}{M_2}\mathbf{v} \times \mathbf{M}_3\right)$$

could be derived from already recognized mechanical force laws (and equations of constraint).

If we are successful in this construction, we need no longer consider '$\mathbf{E}$' as designating an "electric field vector" (thus involving an "inference to an unknown [new] entity"); but instead we can consider '$\mathbf{E}$' as shorthand for the expression '$\mathbf{M}_1$', which refers to a specific complex of "known" (familiar mechanical) entities, the strain in an elastic medium, for example.

What I have been describing is, of course, the program of constructing a "mechanical model of the ether," a task which engaged physicists toward the end of the nineteenth century.

This was the attempt to provide what has been called a "mechanical explanation" of light and other electromagnetic phenomena.

I have been writing in the subjunctive mood for several paragraphs, and I have used words such as 'roughly' and 'sort of'. For some readers it will be anticlimactic to report that the construction program failed. Appropriate functions M_1, M_2, M_3, and M_4 could not be found. E, H, c, and ρ resisted all attempts to reduce them to more familiar mechanical parameters, and Maxwell's equations together with the Lorentz force law stood as independent laws of nature.[13] Electrodynamics stood as a distinct science—or, as I prefer to put it, a distinct *branch* of classical mechanics—alongside gravitational theory and "spring theory."

But we shall see in the following chapters that electrodynamics is an uneasy branch of classical mechanics, for it led to the overthrow of Newton's second law of motion itself.

---

[13] See Whittaker (*op. cit.*) for an exposition of what seems an almost amusing array of mechanical gadgetry that marched through the physics journals in the last decades of the nineteenth century, each attempting without success to be "the real M_1, M_2, ..."

Then there were those of exquisite hindsight who, wearying of the endeavor, marshalled arguments of a "philosophic" nature to show how such attempts were doomed to failure from the outset. (This entire episode in the history of science is rich with lessons to be learned for the philosophy of science, but further exploration here would go beyond the purposes of this book.)

5

THE CONCEPT
OF A FRAME
OF REFERENCE

REFERENCE FRAMES AND
NEWTON'S SECOND LAW OF MOTION

We began our study by introducing the concept of a *frame of reference* with respect to which the positions, velocities, and accelerations of massive bodies are determined. It is important to realize that the concept of a frame of reference is not an abstruse invention of physicists used merely to facilitate the mathematical development of a theory. It is a rather concrete and basic necessity when it comes to the application of physical theories in describing, predicting, and explaining the behavior of real bodies in motion. We must actually lay down meter sticks, look at clocks, and such, when it comes to dealing with the objects in the domain of a theory of mechanics. Moreover, the statement of the fundamental laws of a theory is incomplete until we specify the class of reference frames with respect to which these laws are supposed to hold. We shall see an example of this presently.

Now it is clear that there is some arbitrariness in which possible frame of reference we may wish to use. We will get different numbers for the components of the position vector for a given body at a given time depending on the location of the origin of the chosen reference frame and the direction in which the axes point (and depending upon the particular type of coordinate system which we might associate with the chosen frame of reference). But it is also clear that some frames of reference will not do. Let us see why this is so.

Consider a body upon which no forces are acting—for definiteness, a massive body which is not electrically charged, not attached to any springs, with no other massive bodies in its vicinity, and so on. In other words, if we were to consult a "handbook of force laws for various types of situations," containing equations of the form $\mathbf{f} = \mathbf{f}_i$ (various parameters), we would find that the value of $\mathbf{f}_i$ for each $\mathbf{f}_i$ in the handbook is equal to zero for the body in question (or, is indistinguishable from zero given the fineness of our measurements). Now suppose that we find the acceleration of the body in the given situation to be zero also. (It is moving with some constant velocity, perhaps zero.) These determinations have been made with respect to some previously chosen frame of reference (and associated coordinate system). To say that $\mathbf{v} = 0$, for example, is to say that the rate of change of $\mathbf{r}$ with respect to time is $= 0$, and we must have measured $\mathbf{r}$—for example, x, y, and z—for the body over a span of time.

Newton's second law of motion, $\mathbf{F} = m\mathbf{a}$, holds for this body, since both sides of the equation are equal to zero. More accurately (since the measurements have been carried out in a particular reference frame), we should say $\mathbf{F} = m\mathbf{a}$ holds for the given body *in the chosen frame of reference*.

Consider now the following different frame of reference, in which we might have performed the relevant measurements. Its origin is the same as that of the original frame, and the Z-axes coincide, but it is rotating about the Z-axis of the original frame with constant angular velocity. In this reference frame there will still be no forces acting upon the body. (It is still not electrically charged, and so on.) But the body will be accelerating. If its velocity had been zero with respect to the original frame, it will be moving in a circular path about the Z-axis in the new frame; $a \neq 0$ with respect to the new frame.[1] In this new frame of reference $\mathbf{F} = m\mathbf{a}$ does *not* hold for the body in question. We may say that this new frame of reference is not appropriate for carrying out measurements in mechanics, and we see that only some frames of reference are allowable for doing mechanics. Those reference frames with respect to which $\mathbf{F} = m\mathbf{a}$ does not hold are ruled out as somehow "unnatural."

Now the kind of language I have just used may strike some as being unclear or odd. For example, what does it mean to speak of a law of nature as "holding" or "not holding" in different frames of reference? While most physicists (and some philosophers) are not troubled by this kind of talk, I believe there would be a gain in understanding if the ideas under consideration could be expressed in terms of a relatively clearer terminology. For example, cannot we just say that Newton's second law of motion, which claims quite generally that $\mathbf{F} = m\mathbf{a}$, is simply *false*, although extremely useful in some situations (that is, with the proper choice of reference frame)? This formulation, it seems to me, is misleading. There is another way to go about it, and this involves going back to the canonical form of Newton's second law familiar to philosophers of science, which I briefly discussed in the early pages of Chapter 2. There we had

$$(\xi) \, [(m(\xi) = M) \cdot (a(\xi) = A) \cdot$$

$$(\mathrm{F}(\xi) = \mathrm{f}) \cdot \supset \cdot (\mathrm{f} = MA)]. \, . \, . \, (1)$$

---

[1] See the last few pages of Chapter 1, where it was shown that uniform circular motion entails a nonzero acceleration.

It should be emphasized that we have here only one concrete situation, not two, but we are considering two descriptions of it, with respect to two distinct frames of reference.

(with the explanation of this formulation given in Chapter 2). Let us rewrite this more briefly as

$$(\xi) \; [\mathbf{F}(\xi) = m(\xi) \; \mathbf{a}(\xi)] \qquad \ldots (2)$$

Now this expression of Newton's second law is *incomplete*, since it contains no reference to the frame of reference with respect to which $\mathbf{a}(\xi)$ is to be determined. It is, quite literally, *useless* for doing mechanics. We cannot say whether the statement is true *or* false, because the expression $\mathbf{a}(\xi)$— acceleration of the body ξ—is indeterminate until we specify the reference frame with respect to which the acceleration is to be measured. Suppose, then, that we were to rewrite expression (2) with another universal quantifier, ϕ, ranging over frames of reference:

$$(\phi)(\xi) \; [\mathbf{F}\,(\xi, \phi) = m\,(\xi, \phi) \; \mathbf{a}\,(\xi, \phi)] \qquad \ldots (3)$$

In English, "if ϕ is *any* reference frame and ξ is any body, the force acting on ξ in ϕ is equal to the product of the mass of ξ in ϕ and the acceleration of ξ *in* ϕ." But we have seen that this is simply *false*, for (in the previous terminology) $\mathbf{F} = m\mathbf{a}$ does *not* "hold" in *all* frames of reference. We can get an acceptable expression of Newton's second law, which takes into account its relativization to specific reference frames, and avoids terminology such as 'holds', as follows:

We introduce the predicate 'P' to characterize in some general (but as yet unspecified) way the physical characteristics of those frames of reference in which Newton's second law does "hold." The correct and *complete* expression of Newton's second law of motion is then

$$(\phi) \; \{P_\phi \bullet \equiv \bullet (\xi)[\mathbf{F}(\xi, \phi) = m(\xi, \phi) \; \mathbf{a}(\xi, \phi)]\} \quad \ldots (4)$$

In English, "If ϕ is any frame of reference of *type P*, then (and only then) will the force acting on any body ξ in ϕ

be equal to the product of the mass of ξ in ϕ and the acceleration of ξ in ϕ."

We are left, of course, with a large problem. Can we specify P in a noncircular way—without simply saying that P-type frames are those in which Newton's second law holds? I shall postpone discussion of this issue until the end of the chapter.

Let us now go back to the two frames of reference discussed earlier. Call the frame in which $\mathbf{F} = m\mathbf{a}$ "held" 'frame-1', and call the one in which $\mathbf{F} = m\mathbf{a}$ did not "hold" 'frame-2'. That is, 'P' is true of frame-1, or 'P_1' is true. Similarly, 'P_2' is false. Now it is a simple exercise in logic to see that statement (4) and 'P_1' together imply the statement

$$(\xi) [\mathbf{F}(\xi, 1) = m(\xi, 1) \, \mathbf{a}(\xi, 1)]^2 \qquad \ldots . (5)$$

Moreover, since we have agreed to accept both (4) and 'P_1' as being true, we must accept (5) as true. But (5) is just the complete expression of Newton's second law of motion as *instantiated for frame-1*. We have therefore expressed more clearly (if a bit pendantically) the original statement that Newton's second law "holds" in frame-1. Briefly, "If '1' is a P-type frame, then a certain *consequence of* Newton's second *law* (that *law* being accurately and completely expressed by (4)) is true." (The *consequence* in question is expressed in (5). It *resembles* Newton's second law, but it is not identical with it.[3])

Consider now frame-2, in which "Newton's second law

---

[2] Note that '1' appears wherever 'ϕ' appeared in (4). So we have deduced "for any body ξ the force on ξ in frame-1 is equal to the product of the mass of ξ in frame-1 and the acceleration of ξ in frame-1."

[3] It is not even a "law of nature" according to certain philosophical analyses of this concept, for it refers essentially to a specified object: frame-1. (Thus it is not completely general.)

does not hold." We know that 'P_2' is false; that is, we accept '$\sim P_2$' ("not P_2") as true. But statement (4), Newton's second law of motion, and '$\sim P_2$' together imply the statement

$$\sim (\xi) \, [\mathbf{F}(\xi, 2) = m(\xi, 2) \, \mathrm{a}(\xi, 2)]^4 \qquad \dots \, (6)$$

This expresses more clearly the original statement that Newton's second law does *not* "hold" in frame-2.

We have seen, then, that before Newton's second law, as expressed by (4), can be of any use in a particular problem, we must have a P-type frame in which to make the relevant measurements.[5] If we have such a frame of reference, we can suppress explicit reference to it in general discussions, and simply use '$\mathbf{F} = m\mathrm{a}$' as the correct expression of Newton's second law, with the aside that $\mathbf{F} = m\mathrm{a}$ does "hold" in the frame being used (where the 'hold' terminology is analyzed as above).

THE CONCEPT OF A CHANGE OF REFERENCE FRAME— THE GALILEAN PRINCIPLE OF RELATIVITY

Let us now suppose that we do have such a frame, a frame to which the as yet unspecified 'P' applies. Let 'K' be the

---

[4] '2' appears wherever 'ϕ' appeared in (4). So we have deduced "it is *not* the case that for any body . . ." where '. . .' is as in footnote 2.

Note that (6) is *not* the same as

$$(\xi) \, [\mathbf{F} \, (\xi, 2) \neq m \, (\xi, 2) \, \mathrm{a} \, (\xi, 2)]$$

which says that $\mathbf{F}$ is *never* equal to $m\mathrm{a}$ in frame-2.

[5] If we do not, we are led into questionable concepts such as that of "centrifugal force," which a number of physics textbooks still use (sometimes with apologies). These are sometimes called 'fictional forces'. This is perhaps because the parameter interpretability condition is not satisfied for force laws containing such forces.

name of it. We shall call such a frame an *inertial frame*. Suppose further that we have another frame which is moving along the X-axis of K at uniform velocity u with respect to K, as shown below. Call this frame 'K''.

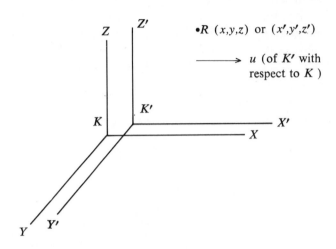

If at time t the position R of a body *with respect to K* is (x, y, z), its position *as measured in K'*, (x', y', z'), will be related to (x, y, z) as follows:

$$x' = x - ut$$
$$y' = y; z' = z$$

And we can add, as an indication that we can use the same clocks in each frame,

$$t' = t$$

Let us give this set of equations the name (for future purposes) *Galilean Transformation*. They indicate how measurements in K can be rewritten or transformed into measurements in K' (and vice versa). (The reader should pause long enough to get an intuitive feel for the equations.) These equations appear to be too obvious to require any explanation or justification. It may come as a surprise to some readers to be told that they will be rejected later. But we shall simply accept them for the present as an essential part of *classical kinematics*. (Recall the penultimate paragraph of Chapter 1.)

By hypothesis Newton's second law "holds in K." Does it also hold in K'? That is, is $\mathbf{F}' = m'\mathbf{a}'$ true (where the "prime" indicates measurements with respect to K') if $\mathbf{F} = m\mathbf{a}$ is true (where the unprimed '$\mathbf{a}$' etc. indicate measurements with respect to K)? Let us find out. '$\mathbf{F} = m\mathbf{a}$' can be written as

$$\mathbf{F} = m\frac{d^2\mathbf{r}}{dt^2}$$

In K this encapsulates three scalar equations:

$$F_x = m\frac{d^2x}{dt^2}, F_y = m\frac{d^2y}{dt^2}, F_z = m\frac{d^2z}{dt^2}$$

We want to express the information contained in these equations *with respect to K'*. Using the Galilean transformation equations, we get

$$\frac{dx'}{dt'} = \frac{d}{dt}(x - ut), \text{ or } \frac{dx'}{dt'} = \frac{dx}{dt} - u$$

$$\frac{dy'}{dt'} = \frac{dy}{dt}; \frac{dz'}{dt'} = \frac{dz}{dt}$$

Further differentiation yields

$$\frac{d^2x'}{dt'^2} = \frac{d}{dt}\left(\frac{dx}{dt} - u\right), \text{ or } \frac{d^2x'}{dt'^2} = \frac{d^2x}{dt^2}$$

$$\frac{d^2y'}{dt'^2} = \frac{d^2y}{dt^2}; \frac{d^2z'}{dt'^2} = \frac{d^2z}{dt^2}$$

Going back to the vector form (and assuming $\mathbf{F}' = \mathbf{F}$ and $m' = m$), we have

$$\mathbf{F}' = m'\mathbf{a}'$$

Newton's second law holds in K' also. The effect of a force acting on a body, when the effects are measured in K', is once again an *acceleration, with respect to K',* such that $\mathbf{F}' = m'\mathbf{a}'$.

The actual *components* of the position and force vectors may be quite *different* in K and K', but they *vary together* in such a way that the *form* of the *vector relationship* remains the same. This is usually summed up by saying that $\mathbf{F} = m\mathbf{a}$ *is invariant under a Galilean transformation.*

We have shown that if $\mathbf{F} = m\mathbf{a}$ holds in K—if K is a P-type frame—then it holds in K'. If K is an inertial frame, any frame moving uniformly with respect to K is also an inertial frame. The clearest way to express this result is

If K and K' are any two frames of reference moving uniformly with respect to each other, then K is an inertial frame if and only if K' is, where 'inertial frame' means '$\mathbf{F} = m\mathbf{a}$' holds.

This statement is known as the *Galilean Principle of Relativity.* (It presupposes the Galilean transformation equations as the appropriate mode of translating measurements between frames moving uniformly with respect to each other.)[6] It is important to notice that this principle has nothing to do with Einstein's theory of relativity. The principle is a part of *classical* dynamics which was recognized in the nineteenth century—in fact Newton at least implicitly was aware of it and accepted it.

There are a number of other formulations of this principle that have appeared in books on mechanics, both pop-

---

[6] The Galilean principle of relativity, as I have stated it, is an analytic consequence of the Galilean transformation equations.

ular and technical. They are not, unfortunately, equivalent to one another. Some are misleading, though highly suggestive; others are corollaries to our statement of the principle. Here are a few, chosen at random:

a) All inertial frames are equivalent for doing mechanics.
b) Absolute velocity cannot be detected by mechanical experiments.
c) There is no reference frame which is absolutely at rest, as far as mechanics is concerned.
d) Velocity is an essentially relational concept in mechanics.
e) If K is moving uniformly with respect to K', no mechanical experiment will determine which is "really" moving.
f) If absolute space does exist, mechanical experiments will not enable us to detect it.[7]

Formulation a) emphasizes the fact that if K and K' are both inertial frames, it does not matter which is used in making measurements in mechanics. Description, explanation, and prediction can equally well be carried out in each frame.

Formulations b) through f) emphasize, each in its own way, the fact that when we say that the velocity of a body is so-and-so, we have said nothing unless we specify the reference frame with respect to which this velocity is determined. But c), for example, can be misleading; it may suggest that all frames of reference are in motion. (With respect to *what*?) And b) suggests that there *is* something, namely absolute velocity, which we unfortunately cannot detect.

Finally, these formulations, by using the term 'mechanical' in an uncritical manner, neglect the very important fact

---

[7] These formulations all rest on the fact that the constant relative velocity of K and K', u, dropped out upon double differentiation of the equation $x' = x - ut$.

that invariance has been shown only for $\mathbf{F} = m\mathbf{a}$. In our statement of the principle no mention has been made of the various *specific force laws* that are used in conjunction with $\mathbf{F} = m\mathbf{a}$ in explaining the motions of bodies. We must also ask whether those specific force laws are invariant under a Galilean transformation. For if they are not, then different inertial frames *could* be distinguished by mechanical experiments! In demonstrating the invariance of $\mathbf{F} = m\mathbf{a}$ under a Galilean transformation, after all, it has been shown only that the *acceleration*, a, of a body is the same in two frames moving uniformly with respect to each other. And this result is significant only if the specific *force* functions that are put into the left side of $\mathbf{F} = m\mathbf{a}$ are also invariant under a Galilean transformation.[8] In brief, formulations a) through f) *do not follow* from our formulation of the Galilean principle of relativity, but follow only from that principle in conjunction with the further assumption that all known specific force laws are also invariant under a Galilean transformation. One exception would suffice to falsify formulations a) through f), for it would provide a mechanical means for distinguishing among inertial frames.

Now it turns out that *most* specific force functions discussed in earlier chapters are invariant under a Galilean transformation. Consider, for example, Hooke's law, $f^s = -ks$, where 's' represents the displacement of a body attached to a spring from the untensioned, or "normal," position. 's' represents the difference between two positions (measured with respect to a given reference frame K). If we let x_1 be the normal position and x_2 the displaced position, Hooke's law can be written as

$$f^s = - k \, (x_2 - x_1)$$

---

[8] We have been assuming throughout that the *mass* of a body is the same in all reference frames. This assumption is a tacit part of classical dynamics.

If we translate this law into measurements with respect to another frame K' (as described above) with the aid of the Galilean transformation equations, we get

$$x_2' = x_2 - ut$$
$$x_1' = x_1 - ut$$

so that

$$(x_2' - x_1') = (x_2 - ut) - (x_1 - ut) = (x_2 - x_1)$$

Thus, in K', $f^s = - k(x_2' - x_1') = - k(x_2 - x_1)$. Hooke's law does not change in form.

In a similar fashion, invariance can be demonstrated for Newton's law of gravitation and all other force laws of mechanics—*but for the Lorentz force law*! It will be shown in the following chapter that the Lorentz force law is *not* invariant under a Galilean transformation. This one exception is of utmost significance, as might be surmised from my remarks about formulations a) through f) above. It initiates, as we shall see, the downfall of classical mechanics.

It has become clear, then, that the concept of a *frame of reference* in which laws are expressed and measurements made is more than a logical nicety required for the neat exposition of a theory, and more than a boring practical necessity in the application of a theory. A careful consideration of this concept forms an essential part of the Einsteinian revolution.

There is one piece of unfinished business to attend to. What is there about inertial frames of reference which distinguish them from frames in which $\mathbf{F} = ma$ does not hold? What is the 'P' in equation (4)? So far, all we know is that *if* K is an inertial frame (a P-type frame), then so are all other frames moving uniformly with respect to K. We have stated how inertial frames are related to each other, but what do they have in common? Within classical mechanics there is no answer to this question! The best we can do is to say that *if* we do happen to run across a frame in which $F = ma$ holds, we *then* know how to generate an infinite class of other such frames. But there is no physical procedure that can be read out of the laws of mechanics which will tell one *how to find* a frame in which $F = ma$ holds.

We do, however, know the following *brute fact*: a frame of reference which is moving uniformly with respect to the distant galaxies is an inertial frame.[9] But within classical (or even special relativistic) mechanics there is no explanation of this fact. We may use this cosmic coincidence to express *P*, but the formulation of this coincidence would be a mere empirical correlation with no theoretical underpinning. This is achieved only within the context of the Theory of *General* Relativity, which is beyond the scope of this book. (See Chapter 12 for a further discussion of the concept of an inertial frame, and its relation to Newton's laws of motion.)

[9] The expansion of the universe calls for a more accurate statement of this, but the formulation in the text is sufficient for present purposes.

6

REFERENCE FRAMES
AND
ELECTRODYNAMICS

ELECTRODYNAMICS AND
THE GALILEAN TRANSFORMATION

We know that Newton's second law of motion, as well as the various specific force laws (which together constitute the theoretical structure of classical or Newtonian dynamics), are invariant under a Galilean transformation—but for the Lorentz force law. If we are fortunate enough to have a frame for which $\mathbf{F} = m\mathbf{a}$ holds,[1] it is sufficient for formulating and applying the laws of classical mechanics, no matter how practically awkward or "unnatural" this frame might otherwise appear to be[2]—again, except for the Lorentz force law. We must now examine this crucially important exception.

Suppose that we have an inertial frame K in which the Lorentz force law

$$\mathbf{f^{EH}} = e\left(E + \frac{1}{c}\mathbf{v} \times \mathbf{H}\right) \qquad \ldots (1)$$

---

[1] Where this terminology ('holds') is to be analyzed as in the previous chapter.

[2] For example, if $\mathbf{F} = m\mathbf{a}$ holds in a reference frame fixed in a high-speed train, the train is theoretically sufficient for doing mechanics. It is no worse than any more usual frame attached to the earth.

holds. This vector equation encapsulates three scalar equations relating the components of the vectors (with respect to K) to each other. Consider now another reference frame K' moving along the positive X-axis of K with uniform velocity u.[3] Will the Lorentz force law preserve its form when translated into measurements with respect to K' in accordance with the Galilean transformation equations? The answer is No, as should be apparent from the occurrence of the term 'v' in eq. (1).[4] The first Galilean transformation equation

$$x' = x - ut$$

upon differentiation yields

$$\frac{dx'}{dt} = \frac{dx}{dt} - u$$

or

$$\mathbf{v}' = \mathbf{v} - u$$

The second term in the Lorentz force law, containing as it does a velocity term, will therefore be *different* in K' than it is in K. This can also be expressed by saying that if

$$\mathbf{f}^{\text{EH}} = e\left(\mathbf{E} + \frac{1}{c}\mathbf{v} \times \mathbf{H}\right)$$

holds in K, then

$$\mathbf{f}^{\text{E'H'}} = e'\left(\mathbf{E}' + \frac{1}{c'}\mathbf{v}' \times \mathbf{H}'\right)$$

---

[3] Do not confuse v with u. The former represents the velocity of a charged body with respect to K; the latter represents the velocity of K' with respect to K: it is a constant relating the two frames to one another.

[4] The Lorentz force law, unlike all other specific force laws of classical mechanics, contains a term expressing the velocity of the body under investigation *with respect to the frame of reference*. Hooke's law with damping also contains a velocity term, but it is the velocity of the body under investigation *with respect to the resist-*

will *not* hold in K'. In brief, *not* all inertial frames are equivalent for doing that *branch* of classical mechanics known as electrodynamics. It will be useful to see how this nonequivalence is to be spelled out in detail. For this purpose it is more enlightening to examine the equations of constraint (Maxwell's equations) relating the parameters which occur in the Lorentz force law rather than the Lorentz force law itself. It will be seen that Maxwell's equations too are not invariant under a Galilean transformation, and we shall examine the consequences of this noninvariance.

Maxwell's equations contain the parameters e (or ρ), c, v, E, and H. We must use the Galilean transformation equations to obtain e' (or ρ'), c', v', E', and H': that is, the values of these parameters with respect to K'. But this procedure is mathematically cumbersome, and so I shall instead examine the transformation characteristics of one of the *consequences* of Maxwell's equations: the *wave equation*.[5]

It was shown in Chapter 4 that Maxwell's equations allow for traveling electromagnetic waves. The space-time behavior of such waves can be expressed, as we saw, by equations such as

$$E = e^{i(\kappa x - \omega t)} \qquad \qquad \ldots (3)$$

---

ing medium, that is, with respect to *other bodies* in the environment. As such, it represents the first derivative with respect to time of a *difference* of positions, and does *not* change in value under a Galilean transformation. (Later in this chapter I shall argue that this apparent difference in the two cases is illusory.)

[5] This is legitimate, for if this consequence of Maxwell's equations is not invariant under a Galilean transformation, it follows that Maxwell's equations themselves are not invariant either.

Moreover, the physical significance of the noninvariance of Maxwell's equations stands out more clearly in the case of this particular consequence of them, as we shall see presently.

where $\omega/\kappa = c$ (with $\omega = T/2\pi$, and $\kappa = \lambda/2\pi$). Suppose that eq. (3) holds in inertial frame K. Does it also hold in K'? We use the first of the Galilean transformation equations

$$x' = x - ut$$

to reexpress eq. (3) with respect to K', obtaining

$$\mathbf{E}' = e^{i[\kappa(x'+ut)-\omega t]}$$

which can be rewritten as

$$\mathbf{E}' = e^{i[\kappa x'-(\omega-\kappa u)t]} \qquad \ldots (4)$$

Equation (4) looks like eq. (3), but for the fact that $(\omega - \kappa u)$ multiplies t rather than ω alone. We can emphasize this by introducing ω' as shorthand for $(\omega - \kappa u)$ to get

$$\mathbf{E}' = e^{i(\kappa x'-\omega' t)} \qquad \ldots (4')$$

Now just as eq. (3) represents a wave moving along the positive X-axis with velocity $c = \omega/\kappa$, so eq. (4') represents a wave moving along the positive X'-axis with velocity $c' = \omega'/\kappa$, where the velocities of the wave in the two frames are related by

$$c' = \frac{\omega'}{\kappa} = \frac{\omega - \kappa u}{\kappa} = \frac{\omega}{\kappa} - u = c - u$$

The velocity of the electromagnetic wave with respect to K' is less than its velocity as measured in K by the quantity u, the relative velocity of the two frames.[6] The original wave equation, expressed *in K* by

$$\frac{\partial^2 E_x}{\partial x^2} - \frac{1}{c^2}\frac{\partial^2 E_x}{\partial t^2} = 0 \quad \ldots \text{ (Ch. 4, eq. (6))}$$

[6] The period will also be different, $2\pi/(\omega - \kappa u)$, but the wavelength will be the same in the two frames.

will take the form

$$\frac{\partial^2 E'_{x'}}{\partial x'^2} - \frac{1}{(c-u)^2}\frac{\partial^2 E'_{x'}}{\partial t^2} = 0$$

in frame K'. We can now see that Maxwell's equations, from which the wave equation is derived, will also change their form when transformed from K to K'. The factor u, expressing the relative velocity of K and K', will appear in the appropriate places in Maxwell's equations. In brief, *Maxwell's equations are not invariant under a Galilean transformation between inertial frames.*

From the noninvariance of Maxwell's equations and of the Lorentz force law we may conclude that *not all inertial frames are equivalent for doing that branch of classical mechanics known as electrodynamics.*

The specific form of this nonequivalence (or noninvariance) is important. Each inertial frame of reference—each frame in which $\mathbf{F} = m\mathbf{a}$ holds—is distinguished from all other such frames by a characteristic constant: *the velocity of electromagnetic waves with respect to that frame.* And one particular inertial frame[7] stands out as *unique.* In that frame the velocity of electromagnetic waves is the same in all directions: Maxwell's equations with 'c' hold in that frame. In all other frames the constant '$c - u$' occurs in place of 'c,' and the velocity of electromagnetic waves will be different in different directions.[8]

Now, it is essential at this point to realize that none of the results discussed so far should come as a surprise. For

---

[7] Strictly speaking, one particular *class* of inertial frames (at rest with respect to each other).

[8] At least this is what the theory predicts. As we shall see in the following chapter, experiment seems to tell another story. (If experiment had turned out to agree with these predicted consequences of theory, we might still be doing classical mechanics today.)

if it is assumed that electromagnetic waves are traveling disturbances in an underlying elastic medium, an "ether," then the assumption that the wave equation (with the constant c) holds in K amounts to assuming the K *is at rest with respect to this ether.* In a frame K' which is moving with respect to K, *moving with respect to the ether*, we should *expect* Maxwell's equation to assume a different form, with a new constant, $c - u$, reflecting the velocity of K' with respect to the ether.

The point is important enough to merit emphasis with the aid of the following analogy: Consider Hooke's law with damping, $f = -kx - cv$, which we assume holds in a given frame K which is at rest with respect to the ambient viscous medium. If we use the Galilean transformation equations to rewrite this law with respect to a frame K' moving with respect to K (with velocity u along the positive X-axis), it will take the form $f = -kx' - c(v - u)$. We *could* say that Hooke's law with damping is *not* invariant, and the result would not be startling. After all, we should expect uniqueness for the frame which is at rest with respect to the viscous medium in which our apparatus is immersed. We normally *avoid* the conclusion of noninvariance in this case by specifying that 'v' is to refer to the velocity of the body with respect to the medium, not with respect to just *any* frame.

In a similar way, we could avoid ascribing noninvariance to the laws of electrodynamics by specifying that the terms 'v' and 'c' which occur in these laws are to refer to measurements with respect to the *ether* (that is, with respect to a frame which is at rest with respect to the ether). This may appear odd to some readers, for we now seem to be concluding that the laws of electrodynamics *are* invariant under a Galilean transformation after all. But once it is assumed that electromagnetic phenomena occur in some sort of underlying quasi-material medium, the parallel with the damping case is inescapable. We may describe both cases in either of two ways: (1) The respective laws are *invariant* under a Galilean transformation *if* we take into account reference to the background medium in which the physical processes occur, or (2) The respective laws are *not* invariant under a Galilean transformation, and they allow the selection of *privileged* inertial frames; but we should not have expected otherwise. It seems to me to be a matter of taste as to which of these formulations is to be preferred.[9] In what follows I shall use the second formulation when discussing electrodynamics, since it is more in accord with the "standard" presentations of the theory. In capsule form, then, the important conclusion we have reached so far in this chapter is *The laws of classical electrodynamics are not invariant under a Galilean transformation between inertial frames of reference.*

THE ETHER FRAME

Earlier we introduced the concept of an *inertial frame* as one in which $\mathbf{F} = m a$ holds. In a similar fashion we may speak of a *Maxwellian frame* as one in which the laws of

[9] I have dwelt on this at some length because many treatments of the transformation properties of the laws of electrodynamics simply "demonstrate" the noninvariance of these laws under a Galilean transformation and move on, leaving the reader with the impression that certain conceptual difficulties in classical physics have *thereby* been unearthed.

There are enough such difficulties (as we shall see) to help generate relativistic mechanics without manufacturing nonexistent ones.

Perhaps the briefest way of putting my present point is: *There is nothing inherently wrong with noninvariance as such.* In the case of electrodynamics the noninvariance can be *explained* if an ether is assumed (just as in the case of the viscous medium and Hooke's law with damping).

classical electrodynamics hold. With the aid of this terminology our results so far can be formulated as follows:

If K and K' are two *inertial* frames (moving uniformly with respect to each other), then: if K is also a *Maxwellian* frame, then K' is not.

This has sometimes been expressed by saying that *the Galilean principle of relativity does not hold*[10] *for electrodynamics.* If we were now to identify the *ether* with *absolute space*, we could express this result by taking over the other formulations of the Galilean principle of relativity from Chapter 5 in their *negative* form. We would then obtain

a) Not all inertial frames are equivalent for doing *electrodynamics.*

b) Absolute velocity *can* be detected by *electrodynamic* experiments.

c) There *is* a reference frame which is absolutely at rest, and it can be found by doing the appropriate *electrodynamic* experiments.

d) Velocity is no longer a relational concept if we bring in electrodynamics.

e) If K is moving with respect to K', *electrodynamic* experiments will determine which is really moving.

f) If there is absolute space, electrodynamics will enable us to detect it.[11]

Again, these formulations are acceptable *if* the ether is

[10] The "hold" terminology in this context can be analyzed in a manner similar to that used in the preceding chapter. We shall have occasion to do just that in the following chapters.

[11] These formulations all rest on the fact that the relative velocity of K and K', u, does *not* drop out upon *single* differentiation of the equation $x' = x - ut$ (compare Chapter 5, footnote 7).

identified with absolute space. But one *need not* make this identification. To see this consider again what I called the application of Russell's maxim to the electrodynamic parameters: the attempt to reduce these parameters to those which enter into the force laws of preelectrodynamics-physics.[12] We may distinguish historically two stages with regard to this reductive program. First, there was the period during which it was believed that a "mechanical model of the ether" could be constructed and that electrodynamic phenomena were at bottom "mechanical" phenomena (in the restricted sense of preelectrodynamics mechanics). Second, there was the period during which it came to be admitted that the reductive program had failed and that electrodynamic phenomena were *sui generis*.[13]

During the first period there was obviously no *necessity* to identify the ether with absolute space (assuming the latter concept to make sense at all), for this ether was conceived as a (as yet unspecified) mechanical object, or medium, *in* space. In fact, it would be a *mistake* to make this identification. (Why should *this* particular mechanical object rather than any other be the correct candidate for the identification?) Therefore, formulations a) through f) would be incorrect, and phrases like 'absolute space' should be replaced by 'ubiquitous mechanical ether.' (Thus, for example, b) should be rewritten as "Velocity *with respect to the ether* can be detected by electrodynamic experiments.")

During the second period there was also no necessity to identify the ether with absolute space. The failure of the

[12] See Chapter 4.

[13] The distinction I have drawn is highly idealized. During each "period" there were physicists who held the view I have ascribed to the other. But what I am really after here is the *logical* distinction between two *attitudes* which one might take regarding the relation between electrodynamics and the rest of mechanics.

reductive program entails only that electrodynamic entities are not mere complexes of *familiar* mechanical entities, but are "new" sorts of mechanical entities ('mechanical' in the sense of Chapter 3, with the Lorentz force law and its parameters on a par with, rather than reducible to, other force laws and their parameters). Verificationist qualms aside, it is conceivable that the electromagnetic ether, even if it is not a "mechanical" medium in the restricted sense of preelectrodynamics mechanics, could itself move with respect to absolute space. Formulations a) through f) are therefore incorrect in this case also. Once again, they should be rewritten in terms of velocity with respect to the ether (nonmechanical, in this case).

Why have I spent so much time dealing with formulations a) through f), which I consider to be incorrect? The reason is this: In Chapter 5 it was shown that the positive versions of these statements are incorrect formulations of the Galilean principle of relativity. This conclusion is reinforced by showing that the negative versions are incorrect formulations of the fact that the Galilean principle of relativity does *not* hold for the laws of electrodynamics. The point of all of this is to see clearly that *even within classical physics* velocity is always velocity with respect to *other bodies* (or with respect to a *reference frame* fixed to some body). The noninvariance of the laws of electrodynamics does not alter this basic fact.[14] *Even within classical mechanics, including classical electrodynamics, velocity is an essentially relational concept.*

---

[14] This is another case in which some presentations of the theory of relativity manufacture nonexistent conceptual difficulties in classical mechanics, for they often leave the impression that classical electrodynamics, together with the Galilean transformation equations, leave room in classical physics for the concept of "absolute velocity."

The correct conclusion of significance to be drawn from the noninvariance of the laws of electrodynamics under a Galilean transformation between inertial frames is simply this:

Among all inertial frames of reference (which are all equivalent with respect to preelectrodynamics mechanics) there stands out one *unique inertial frame*:[15] that frame which is at rest with respect to the ether (considered as a "mechanical" ether or otherwise), so that the laws of electrodynamics with the constant c hold in that frame. And this fact, deduced from the laws of classical physics, gives rise to the following natural question:

Can we actually find this unique inertial frame? If, as seems reasonable, the earth is moving through the ambient ether, what is our velocity with respect to a frame of reference at rest with respect to the ether—the *ether frame*? In the next chapter we shall examine one classic attempt to determine our velocity with respect to the ether: the Michelson-Morley experiment. Notice that theory guarantees that there is something here to be measured, and it is thus a "mere detail" to actually perform the measurement involved.[16] As we shall see, it was the failure of attempts to supply this detail that gave rise to the kinds of problems that initiated the downfall of classical mechanics.

We may sum up the results so far in terms of two principles and one experimental question:

1) *The Galilean Principle of Relativity*: If K and K' are two frames of reference moving uniformly with respect to

[15] Again, strictly speaking, one *subset* of the set of all inertial frames.

[16] It should be pointed out that what is from the theoretical standpoint a mere experimental detail may require of the experimental physicist a great deal of ingenuity.

each other, then K is an inertial frame if and only if K' is, where 'inertial' means that $\mathbf{F} = m\mathbf{a}$ (and the preelectrodynamics force laws) hold, and where the Galilean transformation equations are the appropriate mode of translating descriptions of physical phenomena between frames moving uniformly with respect to each other.

2) *The "Maxwellian Principle of Absolutivity"*: Not all inertial frames, as defined above, are equivalent for doing that branch of mechanics known as *electrodynamics*. Among all inertial frames there stands out *one* (subset) in which the laws of electrodynamics hold with the constant c.

3) *The Experimental Question*: Can we devise an experiment which will enable us to detect the unique *ether frame* whose existence is guaranteed by principle 2?

7

THEORY vs. EXPERIMENT

THE MICHELSON-MORLEY EXPERIMENT

What is the velocity of the earth with respect to that pervasive medium which is the carrier of electromagnetic phenomena, in particular, electromagnetic waves?

The velocity of *light* waves, which constitute one portion of the spectrum of electromagnetic waves, is c with respect to the ether (or with respect to a frame K at rest in the ether). Consider a light wave moving along the positive X-axis of K, and let us assume that another reference frame K', fixed to the earth with its axes parallel to those of K, is moving in the same direction as the light wave with velocity u. We would expect that the velocity of the light wave is $c - u$ with respect to the earth frame K'. Can this theoretical expectation be experimentally verified? We shall now examine one experiment designed to measure the velocity of the earth with respect to the ether.[1]

---

[1] The experiment I am about to present was first carried out by A. A. Michelson in 1881. I shall describe an idealized version of the "Michelson-Morley experiment." For a description of this classic experiment which shows the ingenuity required to actually set up the equipment to test the expectations of theory see R. S. Shankland, "The Michelson-Morley Experiment," *Scientific American*, Nov., 1964, pp. 107–114. For a review of attempts to repeat this experiment over the following 40 years, as well as a description of other experiments designed to detect the "ether drift," see W. Panofsky and M. Phillips, *Classical Electricity and Magnetism* (Addison-Wesley, 1955), Chapter 14.

A system of mirrors M_0, M_1, and M_2, at rest with respect to K', is arranged as shown below. M_0 is half-silvered, so that part of the light wave is transmitted and part of it is reflected. The light from the source S is thereby split, part of it going out parallel to the "ether drift" (to M_2) and part across the direction of the ether drift (to M_1). When the two beams return, they are reflected from the back of M_0 and directed to a "detection device" D.

u ⟶

assumed velocity of K' (earth and mirrors) with respect to K (ether)

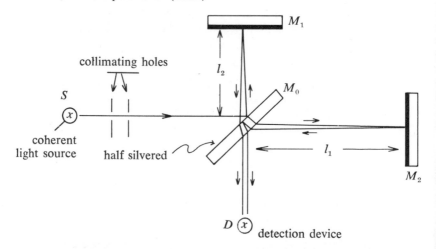

(To simplify the calculations, let us assume that l_1 and l_2 are large in comparison to the distances between M_0 and S, and M_0 and D.[2])

We now ask: How long does it take for that portion of the light beam which traveled *parallel* to the ether drift to

[2] I do this only to make the principle of the experiment stand out more clearly. Obviously, these latter distances could be taken into account in the calculations that follow.

get from S to D? The time elapsed, which I shall designate by '$T_\parallel$,' is equal to the *distance* traveled divided by the *velocity* of the light beam. As measured in the *earth frame K'*, the velocity of the beam as it travels toward M_2 is supposedly $c - u$, and its velocity on the return trip is $c + u$. Therefore, $T_\parallel$ is given by

$$T_\parallel = \frac{l_1}{c - u} + \frac{l_1}{c + u} = \frac{2l_1}{c(1 - u^2/c^2)} \qquad \cdots (1)$$

Similarly, how long does it take for that portion of the light beam which traveled *across* the direction of the ether drift to get from S to D? The path of this beam, *with respect to the ether frame K*, is as shown below. (The lengths are as shown, with 't' designating the time required for the beam to travel between M_0 and M_1.)

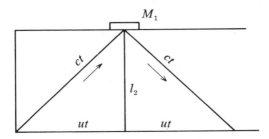

If we let '$T_\perp$' designate the elapsed time in question, then $T_\perp = 2t$. But t can be calculated from the figure as follows:

$$c^2 t^2 = l_2^2 + u^2 t^2$$

Solving for t,

$$t = \frac{l_2}{c} \frac{1}{\sqrt{1 - u^2/c^2}}$$

Finally,

$$T_\perp = 2t = \frac{2l_2}{c} \frac{1}{\sqrt{1 - u^2/c^2}} \qquad \cdots (2)$$

(Notice that if the earth were at rest with respect to the ether, if $u = 0$, we would get from eqs. (1) and (2)

$$T_{\parallel} = \frac{2l_1}{c} \text{ and } T_{\perp} = \frac{2l_2}{c}$$

Moreover, if $l_1 = l_2$, then $T_{\parallel} = T_{\perp}$, as might be expected. But if $u \neq 0$, then $T_{\parallel} \neq T_{\perp}$, even if $l_1 = l_2$.)

If we could measure the ratio of T_n and $T_{\perp}$, we could calculate back to u. The result of the calculation, using eqs. (1) and (2), is[3]

$$u = c \sqrt{1 - \left(\frac{T_{\parallel}}{T_{\perp}}\right)^2 \left(\frac{l_1}{l_2}\right)^2} \qquad \dots (3)$$

(In practice the time ratio itself is not measured, but is calculated on the basis of other directly observable features of the experimental arrangement, together with certain optical laws.[4])

This experiment—or rather a very sophisticated version of it based on the principles described—was carried out a number of times over a period of years, and each time the result obtained was that $u = 0$. The same result was obtained from other experiments designed to measure the velocity of the earth frame K' with respect to the ether frame K. In other words, the wave equation with c rather than $c - u$ seems to hold not only in the supposed ether frame but also in a frame at rest with respect to the earth! Now unless we are willing to assume that the earth is at

---

[3] Eq. (3) can be rewritten as

$$\frac{u^2}{c^2} = \left[1 - \left(\frac{T_{\perp}}{T_{\parallel}}\right)^2 \left(\frac{l_1}{l_2}\right)^2\right]$$

In this form the expression shows that the experiment is capable of detecting what are known as "second order effects," for u^2/c^2 is smaller than u/c if $u < c$.

[4] See the references cited in footnote 1 for details.

rest with respect to the ether (a rather heroic assumption),[5] the experiment seems to show that the laws of electrodynamics hold in *all* inertial frames, contrary to what theory predicts—contrary to what I called the "Maxwellian Principle of Absolutivity" at the end of the last chapter.

THEORY VS. EXPERIMENT

We must pause to absorb the full force of this startling result. For this purpose it will be useful to state the situation in a number of different ways.

The *Galilean Principle of Relativity* said the following:

If K and K' are two frames moving uniformly with respect to each other, then K is an inertial frame if and only if K' is, where 'inertial' means that $\mathbf{F} = m\mathbf{a}$ (and the pre-electrodynamics force laws) hold, and where the Galilean transformation equations are the appropriate mode of translating descriptions of physical phenomena between K and K'.

This can be stated more briefly as[6]

> *The laws of preelectrodynamics mechanics are invariant under a GT between frames moving uniformly with respect to each other.* . . . (4)

---

[5] Notice that it would be required to assume that the earth is at rest with respect to the ambient ether at all times of the year—at all points of the earth's orbit about the sun. This comes dangerously close to assuming that Ptolemy was right after all. (This "possibility" could be tested by carrying out the Michelson-Morley experiment on the moon.) Another way out would be to assume that the earth carries along a portion of the ether as it moves around the sun. But different versions of this "ether drag" assumption fail for reasons which I shall not go into here.

[6] I shall henceforth use '*GT*' as an abbreviation for 'Galilean transformation' and for 'Galilean transformation equations'. (Context will indicate which.)

Now replace 'preelectrodynamics mechanics' in (4) with a *variable* '*T*' ranging over theories to obtain the following statement-*schema*:

> *The laws of T are invariant under a GT*
> *between frames moving uniformly with*
> *respect to each other.*   . . . (5)

Then (4) is but one possible *instance* of (5), a *true* instance. Or, we can say that principle (5) "holds" for preelectrodynamics mechanics.

We can now summarize the results so far as follows:

Theory implies the following:

(a) Principle (5) holds when '*T*' is instantiated by 'preelectrodynamics mechanics.'

(b) Principle (5) does *not* hold when '*T*' is instantiated by 'electrodynamics.'[7]

Experiment seems to show the following:

(c) Principle (5) holds when '*T*' is instantiated by 'preelectrodynamics mechanics.' That is, theory and experiment agree on this point. (Or, at least, no one suspected otherwise at the turn of the century.)

However, (d) Principle (5) also holds when '*T*' is instantiated by 'electrodynamics.'

The Michelson-Morley experiment, encapsulated in (d), contradicts the theoretical expectations stated in (b).

Another (equivalent) way of expressing the situation is this:

Theory implies the following:

(b′) The laws of electrodynamics, *conjoined with GT*, *imply* different c's in different inertial frames. This is to be

---

[7] (b) should be recognized as another expression of the "Maxwellian Principle of Absolutivity." (Chapter 6)

expected given an ether in which light (= electromagnetic) waves are propagated.

Experiment seems to show the following:
(d′) The null result of the Michelson-Morley experiment, *conjoined with GT, imply* the same *c* in different inertial frames, contrary to the expectations expressed in (b′).

It would seem, then, that the laws of electrodynamics are simply false, since what they predict is not observed. However, the logic of the situation is not that simple. Maxwellian electrodynamics had been so well-confirmed, in a variety of experimental situations, that most physicists at the end of the last century could not bring themselves to draw this conclusion. Quite generally, when the expectations of theory disagree with what observation seems to show, various lines of approach are available short of simply concluding that the theory under test has been falsified. These lines of approach fall into two categories (which are *not* mutually exclusive). One can reinterpret the observations and/or modify the auxiliary assumptions (perhaps previously implicit) which were used in conjunction with the theory under test to deduce the specific observations in question.

After the early unsuccessful attempts to detect and measure the "ether drift" a variety of such proposals, some of them rather ingenious, were offered to "explain away" the null results of these experiments. I shall not discuss these proposals here. Suffice it to say that they all failed for various reasons.[8]

---

[8] The "ether drag" hypothesis alluded to in footnote 5 was one such proposal. By far the most interesting attempt to account for the disagreement between theory and experiment was due to Lorentz. He proposed that an object moving with respect to the ether contracted in length by a factor which is a function of its velocity,

REPLACING THE
GALILEAN TRANSFORMATION EQUATIONS

There is another approach one might try. We note that it has been assumed without question that in order to translate descriptions of physical phenomena from one inertial frame to another, the proper way of effecting the translation is via *GT*. These equations were introduced in Chapter 5 as an explicit expression of what appears to be "common sense," and have been used subsequently as if they were totally unproblematic. They appear in the statement of the Galilean principle of relativity expressed in (4) a few pages back, as well as in statements (b') and (d'). We shall find that the way out of the present difficulty lies partly in the replacement of *GT* with other transformation equations. Let us see how this comes about.

Consider again statement-schema (5) and replace '*GT*' with a *variable* '*XT*' ranging over possible sets of transformation equations. This yields the "double" schema:

> *The laws of T are invariant under an*
> *XT between frames moving uniformly*
> *with respect to each other.*   . . . (6)

We now take as a *postulate* the following statement:

> *With the proper instantiation for XT, (6)*
> *must hold for* all *laws of physics (for all*

---

a function which would be exactly right for explaining why the velocity of the earth (and the Michelson-Morley apparatus) with respect to the ether only *seemed* to be zero. His hypothesis, moreover, was not *ad hoc*, for he was able to *deduce* the contraction from his theory of matter, which assumed the *electrical* constitution of all material bodies.

One might also conclude that the experiment shows that light does not consist of electromagnetic waves after all. But no one seriously considered this possibility.

instantiations of T), including
electrodynamics.   . . . (7)

This statement has been called the *Special (or Restricted)*
Principle of Relativity. 'Special' (or 'Restricted') means we
are only considering frames of reference moving *uniformly*
with respect to each other.[9] This principle says that no statement will be considered a law of nature unless it is invariant when transformed between frames moving uniformly
with respect to each other. Which transformation equations
are to be used, what is the proper instantiation for *XT*? We
do not stipulate this in advance, but determine the answer
with the aid of the following additional assumption:

The results of the Michelson-Morley experiment are not
to be explained away, as Lorentz and others attempted, but
are to be accepted at face value. That is, we generalize from
the results of this and similar experiments and assume:

The velocity of light is c *in all*
inertial frames.   . . . (8)

If (8) is considered to be a *law* of nature, then it is but a
special case of (7). It has in fact been called "the law of the
constancy of the velocity of light,"[10] and it can be written in
the canonical form of a "law of nature" familiar to philosophers of science as

$(\phi)(\xi)[(\xi$ is light) $(\phi$ is an inertial frame)

$$. \supset .(v(\xi, \phi) = c)] \ . . . (8')$$

[9] Were we to consider frames moving with respect to each other
in any arbitrary manner, we would have a *General* Principle of
Relativity.

[10] See, for example, Einstein, *Relativity* (Crown, 1961, 17th edition), p. 17. (This excellent little book, which first appeared in
1916, is proof that it is *not* the case that a creator is never a good
interpreter or popularizer of his own work.)

In English, "If ξ is a light wave and ϕ is any inertial frame, then the velocity of ξ as measured with respect to ϕ is c."

Now the possibility of light waves can be explained on the basis of electrodynamics (Maxwell's equations), so it seems as if our approach could be restated as follows:

Find a set of transformation equations such that *Maxwell's equations* will be invariant when transformed between inertial frames in accordance with them. For if Maxwell's equations preserve their form, then (8) will follow automatically since the wave equation is a consequence of Maxwell's equations.

I say, we *could* restate the approach in this manner, but it is not necessary, for, as we shall see, (8) alone is sufficient (together with another rather basic and unproblematic assumption) for determining the appropriate transformation equations! This is a rather important point, for it shows that one need make no theoretical assumptions regarding the physical nature of light in order to find the transformation equations which will yield the null results of the ether drift experiments. It will *turn out*, "fortunately," that Maxwell's equations will also be invariant under the transformation equations that preserve the law of the constancy of the velocity of light. (If this were not the case, then principle (7), the Special Principle of Relativity, would have us discard Maxwell's equations and seek a new theory of light.)

What I am suggesting here runs contrary to the "standard" presentations of relativity theory. My claim is this: While (7) and (8) will suffice for deriving the requisite transformation equations (since (8) can be considered a consequence of (7)), (8) by itself is sufficient. Viewed in this way, the Special, or Restricted, Principle of Relativity becomes a bold extrapolation from the needs of the present situation—an extrapolation with far-reaching consequences for areas of physics other than mechanics (including electrodynamics).

Let us call the as-yet-to-be-determined transformation equations 'LT.' Then (7), the Special Principle of Relativity, will read

> *All laws of physics must be invariant*
> under LT.   . . . (7′)

That is, when 'XT' is replaced by 'LT' in (6), the resulting schema will yield true statements for all instantiations for 'T.'

Statement (7′) has also been referred to as the Special Principle of Relativity. It resembles the *Galilean* Principle of Relativity but for two things:

(i) We need LT rather than GT (since the latter does not preserve the law of the constancy of the velocity of light, (8) or (8′)), and

(ii) *all* laws of physics must be invariant when transformed between inertial frames.

We are saying that *some* principle of relativity—of invariance—must be maintained at all costs, even if we have to give up the "common sense" GT in favor of a new set of transformation equations, and even if, as we may discover, some time honored laws of physics must be rejected if they turn out not to be invariant under the new transformation equations. We have a research directive:

"Seek invariance when formulating physical laws, invariance in accordance with LT, since only LT will make (8) invariant."

Let us then finally turn to the task of determining the new transformation equations LT, and an examination of precisely how the Special Principle of Relativity and the law of the constancy of the velocity of light function in determining them.

8

THE LORENTZ
TRANSFORMATION

DERIVATION OF THE
NEW TRANSFORMATION EQUATIONS

We want a set of transformation equations for translating descriptions of physical phenomena in one inertial frame K into descriptions of the same phenomena in another such frame K', moving with respect to K as shown below.

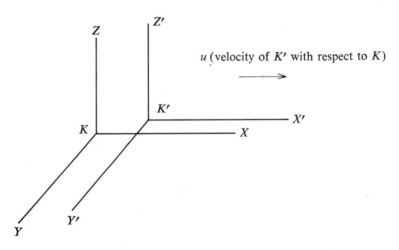

We begin by assuming that the *general form* of the equations should be

$$x' = \alpha x + \beta t \qquad \ldots (1)$$

$$y' = y \qquad \ldots (2)$$

$$z' = z \qquad\qquad \ldots (3)$$

$$t' = \gamma x + \delta t \qquad\qquad \ldots (4)$$

where α, β, γ, and δ are as yet to be determined functions of the velocity u of K with respect to K'. Before determining these functions, let us note a few interesting properties of these equations.

First, equations (2) and (3) state that since the relative motion of the two frames is along their X-axes, there should be no difference in measurements along the Y- and Z-axes.[1]

Second, these equations are *linear*. That is, each primed coordinate is assumed to be a sum of terms involving the unprimed coordinates separately, and each of them only to the first power. There are a number of reasons that can be given for this restriction on the general form of the equations. One of them is this:
Linearity will result in the following additional required feature of the equations:

If u is small compared to the velocity of light, the equations will approach the form of the Galilean transformation equations. That is,

$$\lim_{u/c \to 0} \alpha(u) = 1 \qquad\qquad \lim_{u/c \to 0} \beta(u) = -u;$$

$$\lim_{u/c \to 0} \gamma(u) = 0 \qquad\qquad \lim_{u/c \to 0} \delta(u) = 1$$

This would explain why it had been previously believed that the Galilean transformation equations were correct.[2]

---

[1] There is no loss of generality in assuming that the axes of the two frames form parallel pairs or that the relative motion of the frames is along one of these pairs. I leave it as an "exercise for the reader" to see why this is so.

[2] We could have begun by assuming some more general form for the transformation equations than that given in equations (1) through (4). Linearity could then be deduced. I have chosen to *assume* linearity in order to shorten the presentation.

Finally, there is a fourth equation in which the primed time coordinate involves an unprimed *spatial* coordinate. This feature is not present in the Galilean transformation equations. If it looks uncalled for or strange, notice that at this point it may well turn out to be the case that γ will be equal to 0. If this is so, then we were merely "too cautious" in including this factor, but no "harm" would have been done. However, if this is not so, then we must simply recognize that the resulting equations, as strange as they may look, are required. Let us now determine the functions α, β, γ, and δ.

The origin of $K'(x' = 0)$ is moving with velocity u with respect to K. That is, if there is an object[3] at the origin of K', at rest with respect to K', then $x' = 0$ for this object, and its position with respect to K will be given by $x = ut$. In brief, if $x' = 0$, then $x = ut$. Putting these values into eq. (1), we obtain, for this object,

$$0 = \alpha\, ut + \beta t$$

Solving this equation for β, we get

$$\beta = -\alpha u$$

Therefore eq. (1) can be rewritten as $x' = \alpha x - \alpha ut$, or

$$x' = \alpha(x - ut) \qquad \ldots \text{(a)}$$

We now have only three functions to determine rather than four.

Now eq. (1) tells us (partially) how to translate descriptions in K into descriptions in K'. We can also write down the *inverse* transformation for going from K' to K. Its general form will be the same as eq. (1):

$$x = \alpha'x' + \beta't' \qquad \ldots \text{(1')}$$

where α' and β' are to be determined functions of u. We

---

[3] The term 'object' here refers not only to massive bodies but to *any* identifiable entity which can have a velocity.

now use the same reasoning as above: The origin of K ($x = 0$) is moving with respect to K' with velocity $-u$. That is, if there is an object at the origin of K, at rest with respect to K, then $x = 0$ for this object, and its position with respect to K' will be given by $x' = -ut'$ (since it is moving along the X'-axis with velocity $-u$). In brief, if $x = 0$, then $x' = -ut'$. Putting these values into eq. (1'), we obtain, for this object,

$$0 = \alpha'(-ut') + \beta't'$$

Solving for β',

$$\beta' = \alpha'u$$

Therefore, eq. (1') can be rewritten as $x = \alpha'x' + \alpha'ut$, or

$$x = \alpha'(x' + ut') \qquad \ldots \text{ (b)}$$

If we now combine eq. (a) and eq. (b), and solve for t', we obtain

$$t' = \left(\frac{1 - \alpha'\alpha}{\alpha'u}\right)x + \alpha t \qquad \ldots \text{ (c)}$$

which is a more determinate form of eq. (4).[4] There are now only two functions of u to be determined: α and α'.

We now apply the Special Principle of Relativity to reduce the number of functions to be determined to one. This principle states that all laws of physics must be invariant when transformed between K and K' under the transformation equations we are now seeking. Application of this principle to the case at hand implies that $\alpha = \alpha'$, for otherwise there would be different functions of u involved in

---

[4] Notice that we need not have assumed eq. (4) at the outset, for the preceding argument has just given us a transformation equation for the time coordinate.

equations (a) and (b) above, and this would lead to *different laws* in K and K'. (The expressions for x in terms of x', and x' in terms of x, would be different; therefore the force laws discussed in Chapter 3 would not be the same in K and K', since they involve various functions of the spatial coordinates.)

Now it is essential to notice that *this is nothing new*. A Newtonian physicist, bemused by the strange form of the equations with which we began this chapter, could accept this result, fully expecting that we will eventually end up with the familiar Galilean transformation equations. He could agree that $\alpha = \alpha'$, claiming that they are both equal to 1. In other words, at this point there is no difference between the *Galilean* Principle of Relativity and the *Special* Principle of Relativity. What they have in common *in this case* is their function as a *spatial symmetry principle*. We have required only that space be *isotropic*: that there be no preferred *direction* in space. It does not matter whether we say that K' is moving with respect to K with velocity u or that K is moving with respect to K' with velocity $-u$.[5]

This is what I meant in the last chapter when I said that the Special Principle of Relativity was an extrapolation from the needs of the present situation. This principle *need not* be used in deriving the necessary transformation equations. It does imply the weaker isotropy assumption, which *is* required, but a classical physicist would not object to the latter assumption.

---

[5] If the reader is disturbed by the *apparent* asymmetry of u and $-u$, he should notice that we could have chosen the positive X'-axis of K' as pointing in the opposite direction (thus yielding what is known as a "left-handed" coordinate system associated with this frame of reference). The symmetry of K and K' would then be complete.

So, using the result that $\alpha = \alpha'$, equations (a) and (c) can be written as more determinate forms of equations (1) and (4). We obtain

$$x' = \alpha x - \alpha u t \qquad \ldots (1'), \text{from (a)}$$

$$t' = \left(\frac{1 - \alpha^2}{\alpha u}\right) x + \alpha t \quad \ldots (4'), \text{from (b)}$$

We must, finally, evaluate α. To do this consider an object[3] moving with uniform velocity parallel to the X and X' axes. We ask, What relation does the velocity of this object with respect to K', dx'/dt', bear to its velocity with respect to K, dx/dt?[6] To answer this question, form differentials of equations $(1')$ and $(4')$ above:

$$dx' = \alpha dx - \alpha u\, dt \qquad \ldots \text{from } (1')$$

$$dt' = \alpha dt + \left(\frac{1 - \alpha^2}{\alpha u}\right) dx \quad \ldots \text{from } (4')$$

Divide to get

$$\frac{dx'}{dt'} = \frac{\alpha dx - \alpha u\, dt}{\alpha dt + \left(\dfrac{1 - \alpha^2}{\alpha u}\right) dx}$$

which can be rewritten as

$$\frac{dx'}{dt'} = \frac{\dfrac{dx}{dt} - u}{1 + \left(\dfrac{1 - \alpha^2}{\alpha^2 u}\right) \dfrac{dx}{dt}} \qquad \ldots (5)$$

[6] In order to simplify the ensuing calculations, I shall not consider the general case of an object moving in such a way that the direction of its motion is not parallel with any of the axes. The assumption that only dx'/dt' and dx/dt are nonzero leads to the same result for α as would the more general case.

If we were now to assume, with "common sense," that the velocity of the object with respect to K' is its velocity with respect to K minus the relative velocity of K and K',[7] then eq. (5) would be true only if the denominator on the right were equal to 1, that is, only if $\alpha = 1$. Equations (1) through (4), with which the chapter began, would then be determined in detail; they would in fact be the Galilean transformation equations (GT), for we would have

$$x' = x - ut$$

and

$$t' = t$$

the second equation being a tacit assumption of GT. All of our cautious efforts would have led right back to the familiar GT. But we know that GT is incorrect because it does not preserve the *law of the constancy of the velocity of light*, so that we must reject the classical law of addition of velocities as an aid in determining α. Instead, we apply (8) from Chapter 7: *The velocity of light is c in all inertial frames.* That is, if the "object"[3] we are considering is a light wave (or a light "particle," or *whatever* we may imagine light to be!), its velocity with respect to both K and K' is c. Thus, the application of the law of the constancy of the velocity of light requires that

$$\frac{dx'}{dt'} = \frac{dx}{dt} = c$$

Putting this requirement in eq. (5) yields

$$c = \frac{c - u}{1 + \left(\dfrac{1 - \alpha^2}{\alpha^2 u}\right) c}$$

[7] This assumption is called the "classical law of addition of velocities."

Solving for α, we get

$$\alpha = \frac{1}{\sqrt{1 - u^2/c^2}}$$

and the required transformation equations LT are, finally,

$$x' = \frac{x - ut}{\sqrt{1 - u^2/c^2}}$$

$$y' = y, \quad z' = z$$

$$t' = \frac{t - \dfrac{ux}{c^2}}{\sqrt{1 - u^2/c^2}}$$

These are known as the *Lorentz transformation equations.*[8] They reduce to *GT* as $u/c \to 0$, as required. If we now go back to $(7')$ in the previous chapter, we can express the Special Principle of Relativity as

All laws of physics (mechanics, including electrodynamics—or any other correct theory of light) must be invariant when transformed between inertial frames in accordance with the Lorentz transformation equations.

Two assumptions were used in deriving the Lorentz transformation equations: (1) the Special Principle of Relativity, in a form in which the transformation equations to be used were not explicitly mentioned, and in a case in which it agrees with the Galilean Principle of Relativity; and, more importantly, (2) the law of the constancy of the velocity of light in all inertial frames. It was the latter assumption that played the key role in determining α and thus the exact form of the new transformation equations. The Special

[8] Lorentz originally introduced these equations in his attempt to explain away the null results of ether drift experiments. (See Chapter 7, footnote 8.)

Principle of Relativity thus differs from the Galilean Principle of Relativity in just this respect:

Whereas the Galilean Principle of Relativity simply *noted* that the laws of *pre*electrodynamics mechanics are *in fact* invariant when transformed between inertial frames in accordance with *GT*, the Special Principle of Relativity *requires* that *all* laws of physics should be invariant when transformed between inertial frames. The specific form of the appropriate transformation equations is left open and is determined by the *additional assumption* that the velocity of light is *c* in all inertial frames. Once the specific form of the new equations is determined, the Special Principle of Relativity can be reexpressed (still as a *requirement*) in a form in which the new equations are included in the statement of the principle. In this form the Special Principle of Relativity is quite different from the Galilean Principle of Relativity and has far-reaching consequences for all of physical theory, as we shall see.[9]

SOME PRELIMINARY CONSEQUENCES OF RELATIVITY

I shall conclude this chapter by pointing out some consequences (and nonconsequences!) of the Lorentz transformation equations for classical physics.

(1) In the derivation of *LT no assumptions were made regarding the physical nature of light.* Our results are compatible with light being particles or waves; and if waves, either waves of electromagnetic disturbances or waves due

---

[9] Physical laws (or statements which are candidates for being considered laws) which are invariant under *GT* are not invariant under *LT*, and are *for this reason* (according to the Special Principle of Relativity) to be discarded. An important example of this will be considered in the following chapter.

to some other underlying phenomena. It is thus an open question *at this point* as to whether Maxwellian electrodynamics gives a correct explanation of the behavior of light. In particular, the Special Principle of Relativity requires that the laws of electrodynamics be invariant under *LT* if they are to be considered at all as a possible explanation of the behavior of light (or of anything!).

(2) It turns out ("fortunately," one is tempted to say) that Maxwell's equations are indeed invariant under *LT*. We may still, therefore, identify light with a portion of the spectrum of electromagnetic waves. It is instructive to show this. I shall do so, not for the full set of Maxwell's equations, but for the wave equation. If we write the solution of the wave equation for E_x in the form

$$E_x = e^{i(\kappa x - \omega t)} \quad (\text{where } \frac{\omega}{\kappa} = c)$$

and replace x and t with x' and t' *in accordance with LT* (*not GT*), we shall see that c' (given by ω'/κ') is indeed equal to c, as required if Maxwell's equations are an acceptable account of the behavior of light—in particular, the null results of ether drift experiments!

We get

$$E_{x'} = e^{i\left[\frac{\kappa}{\sqrt{1-u^2/c^2}}(x'+ut') - \frac{\omega}{\sqrt{1-u^2/c^2}}\left(t'+\frac{ux'}{c^2}\right)\right]}$$

This can be rewritten as

$$E_{x'} = e^{i[\kappa'x' - \omega't']}$$

where

$$\kappa' = \frac{1}{\sqrt{1-u^2/c^2}}\left(\kappa - \frac{\omega u}{c^2}\right)$$

and

$$\omega' = \frac{1}{\sqrt{1-u^2/c^2}}(\omega - \kappa u)$$

The velocity of the wave in K', c', is ω'/κ', or

$$c' = \frac{\dfrac{1}{\sqrt{1 - u^2/c^2}}(\omega - \kappa u)}{\dfrac{1}{\sqrt{1 - u^2/c^2}}(\kappa - \dfrac{\omega u}{c^2})}$$

But $\omega = \kappa c$. Putting this into the last equation, we get

$$c' = \frac{\kappa c - \kappa u}{\kappa - \dfrac{\kappa c u}{c^2}}$$

which reduces to

$$c' = c$$

So Maxwell's equations, via the wave equation, do imply that the velocity of light (= electromagnetic waves) is c in all inertial frames, when LT is used to transform between such frames. *The law of the constancy of the velocity of light is a consequence of classical electrodynamics* (when LT is used instead of GT).[10]

If we refer to any theory whose laws are invariant *under LT* as a "relativistic theory," we may sum this up by saying that *classical electrodynamics is a relativistic theory.*[11]

(3) In Chapter 5 an 'inertial frame' was defined as a frame of reference in which $F = ma$ holds. In Chapter 6 I also introduced the notion of a 'Maxwellian frame', as one in which the laws of electrodynamics hold. In the latter chapter we were still assuming that GT is the appropriate set of transformation equations, and we had the result that the laws of electrodynamics single out one inertial frame as unique. That is, we concluded that if K and K' are two

---

[10] See Chapter 7, p. 96.

[11] Electrodynamics is unique among nineteenth-century physical theories in this respect.

inertial frames (moving uniformly with respect to each other), then: if K is also a *Maxwellian* frame, then K' is not. There was a tacit assumption in all of this: Among frames of reference moving uniformly with respect to each other, the one which is Maxwellian *is also inertial*. In theory there was no reason why this should be the case. It is quite conceivable that the class of frames in which $\mathbf{F} = m\mathbf{a}$ holds (all of which are in uniform motion with respect to *each other*) should be executing some arbitrary nonuniform motion with respect to the one Maxwellian ($=$ ether) frame. (The point can be put dramatically, if inaccurately, by saying that the ether could be executing some nonuniform motion with respect to absolute space.)

A similar point can be made now that we are using LT to transform descriptions of physical phenomena between frames moving uniformly with respect to each other. Using LT, we now have a *class* of Maxwellian frame*s* moving uniformly with respect to each other, and there is *no theoretical reason for supposing that the class of Maxwellian frames is identical with the class of inertial frames.*

We shall see in the following chapter that $\mathbf{F} = m\mathbf{a}$ is not invariant under LT and is therefore to be discarded. Newton's first law of motion, however, is invariant under LT,[12] and the concept of an inertial frame can be redefined in terms of the first law. If we do this,[13] the present point can be put as follows: Using LT, there are two basic classes

[12] If the statement 'a body subject to no forces moves with constant velocity in a straight line' holds in a frame K, it will hold in any frame K' moving uniformly with respect to K.

[13] The apparent failure to notice this has led one author to redefine the concept of an inertial frame as one in which Maxwell's equations hold. See M. Bunge, *Foundations of Physics* (Springer-Verlag, 1967), pp. 182–3.

(See Chapter 12 for a further discussion of these issues.)

of reference frames which are central for doing physics: the class of *inertial frames* and the class of *Maxwellian frames*. The members of each class move uniformly with respect to the other members of that class. What is the relation between the two classes? Theory has no answer to this question. Experiment provides us with the following *brute fact*: the two classes are *identical*; a frame of reference is inertial if and only if it is also Maxwellian. The explanation of this empirical fact is related to what I said at the end of Chapter 5 regarding a "*P*-type frame," and is to be found only in the *General* Theory of Relativity.

(4) It is still possible to maintain that electromagnetic waves are disturbances in an underlying medium, or "ether." This is contrary to what has usually been claimed to be a consequence of the Special Principle of Relativity. Most authors claim that the application of a relativity principle to the laws of electrodynamics forces one to discard the concept of an ether and with it a privileged ether frame. But this is not so. It is still conceptually respectable to try to construct a mechanical model of the ether (in the sense of Chapter 4). One need only make sure that the mechanical laws (force laws plus equations of constraint) describing the behavior of this medium are invariant *under LT*. We would not, of course, be able to detect our velocity with respect to this medium, but the requirement of Lorentz-invariance provides a good physical reason for not being able to do so. In other words, the speculations of Lorentz[14] in this regard do not now become "meaningless." Research directed to constructing a mechanical model of the ether was in fact abandoned, but there was *another* good reason for this: namely, the "sociologically sufficient" fact that no one could come up with a general mechanical theory from which

[14] See Chapter 7, footnote 8.

all electromagnetic phenomena (for example, Maxwell's equations) could be deduced! The reductive program failed *in fact*.[15]

The present point is a logical one. Physicists may well have *believed* that the concept of an ether had become "meaningless," but they *need not* have drawn this conclusion. The *worst* thing that can be said about the ether concept in light of relativity is this: The failure of the reductive program, plus the nondetectability of the ether drift, conjoined with a certain positivistic attitude[16] regarding which concepts are "allowable" in physics, *might* lead one to conclude that the ether-concept is *superfluous*. Any stronger conclusion is unwarranted.[17]

[15] The failure led some physicists to attempt the inverse reduction, that is, the reduction of the specific force laws of preelectrodynamics mechanics to the laws of electrodynamics! This met with some success, as some previously known forces were seen to be ultimately due to electromagnetic forces. Today, only four basic kinds of forces are recognized by physicists, electromagnetic forces being one of them.

[16] The attitude I have in mind is in its most general form this: We allow as meaningful only those concepts or statements which can be *individually* "checked" in experience. This attitude has been expressed more specifically in distinct ways by different authors. Einstein rejected it. He wrote:

"In order to be able to consider a logical system as physical theory it is not necessary to demand that all of its assertions can be independently interpreted and 'tested' 'operationally'; *de facto* this has never yet been achieved. In order to be able to consider a theory as a *physical* theory it is only necessary that it implies empirically testable assertions in general."

(See P. A. Schilpp [ed.], *Albert Einstein: Philosopher-Scientist* (Tudor, 1949), p. 679.)

[17] It would not be too misleading to say that within the context of the *General* Theory of Relativity the concept of an ether has returned in a new guise.

(5) Finally, I would like to point out that we could have *postulated LT* as the appropriate manner of transforming between inertial (= Maxwellian) frames rather than *deriving* them from other assumptions. The correctness of *LT* would then be judged on the basis of whether the body of physical theory based on these equations satisfies the usual requirements for acceptable physical theories, such as a high degree of confirmation, simplicity, and so on. But this is just to say that theories can be presented in various ways, all of which are equally "correct." The presentation I have chosen has the advantage of showing how relativity theory can be generated out of the consideration of certain difficulties in classical mechanics including electrodynamics.[18]

Similarly, while the isotropy assumption suffices at one stage of the derivation of *LT*, it would not be "incorrect" to use the full Special Principle of Relativity as the relevant assumption. The latter is logically stronger than the situation requires, and in this sense the Special Principle of Relativity "sticks its neck out" more than is necessary. But it is in fact the very boldness of the Special Principle (as opposed to the weaker isotropy assumption) that led, as we shall see, to a revolution in physics.

---

[18] In this sense my presentation reflects to some degree the history of the development of the theory of relativity, although I have not attempted to give a fully historical presentation.

9

TOWARD A NEW MECHANICS

We now have a new set of equations, the Lorentz transformation equations, LT, which state how one is to translate descriptions of physical phenomena with respect to one frame of reference into descriptions of the same phenomena with respect to another frame moving uniformly with respect to the first. The laws of classical electrodynamics are invariant under this transformation, or, as I said in the last chapter, classical electrodynamics is a relativistic theory.

But we are in trouble. For the other laws of classical mechanics—$F = ma$ and the preelectrodynamics force laws —are *not* invariant when transformed in accordance with the new transformation equations. Clearly, since $F = ma$ preserves its form *under GT*, it cannot preserve its form under some *other* set of transformation equations, in particular, LT. If we have an inertial frame of reference K—that is, a frame in which $F = ma$ holds—then $F = ma$ will not hold in a frame K' moving uniformly with respect to K if we use LT to effect the transformation. Theory predicts that K' will not be an inertial frame, because when accelerations are transformed in accordance with LT, they will change their form—both in magnitude and direction—in a complex way.[1] In brief, if K is inertial, and K' moves uniformly with respect to K, then K' is *not* inertial!

[1] See, for example, R. Katz, *An Introduction to the Special Theory of Relativity* (Van Nostrand, 1964), pp. 48–9, for a simple derivation of how the acceleration vector transforms under LT. (It should be recalled that under GT, $a' = a$. See Chapter 5.)

It now seems that there is a privileged *mechanical*[2] frame K, in which $\mathbf{F} = m\mathbf{a}$ holds, and that we should be able to find this unique frame by *mechanical*[2] experiments. It appears that one could paraphrase what was said in Chapter 6 regarding the "ether frame," with 'electromagnetic ether' replaced by some label such as, say, '*inertial* ether'. The old situation that we had before the Michelson-Morley experiment seems to be coming back to haunt us at a more basic level—the level of the general force law of all mechanics, including electrodynamics—Newton's second law of motion.

THE NEW KINEMATICS

In fact, this noninvariance of the laws of classical mechanics under LT has ramifications even for *kinematics*, the study of the motions of bodies without regard to the forces responsible for their motions. The Lorentz transformation equations yield a new underlying framework within which *dynamical* theories are developed.[3] We have a new, *relativistic kinematics* to replace the old, *classical kinematics*. LT provides an important part of the foundation of the former, while GT plays a similar role in the latter.

What are some of the laws of classical kinematics? Consider again two frames of reference K and K' as described in previous chapters, and suppose that there is a material rod at rest with respect to K' lying along the X'-axis. Its length l', as measured in K', is defined as $x_2' - x_1'$, where x_2' and x_1' are the positions of its end points. What is the length of the rod as measured in K (with respect to

---

[2] In the sense of $\mathbf{F} = m\mathbf{a}$ plus all *pre*electrodynamics force laws.

[3] See the last page of Chapter 1 for the distinction between *kinematics* and *dynamics*.

which the rod is moving at velocity u)? According to GT, $x' = x - ut$ and $t' = t$, so we get

$$l = x_2 - x_1 = (x_2' + ut) - (x_1' + ut) = x_2' - x_1' = l'$$

That is, the length of a rod is a constant: it is the same in all the frames of reference moving uniformly with respect to each other. We may refer to this statement as a "law of classical kinematics," although this bit of "common sense" hardly seems to deserve such an appellation. Similarly, GT implies that temporal intervals (like spatial intervals, or lengths) do not depend on the frame of reference in which they are measured. It is trivial-sounding statements such as these that constitute a (usually implicit) part of classical kinematics. But when we come to use LT as the (correct) mode of translating between K and K', the situation changes dramatically.

Consider again the rod whose length in K' is $l' = x_2' - x_1'$. Using LT to calculate its length with respect to K, we obtain

$$l = x_2 - x_1 = (x_2' \sqrt{1 - u^2/c^2} + ut) - (x_1' \sqrt{1 - u^2/c^2} + ut)$$

or

$$l = (x_2' - x_1') \sqrt{1 - u^2/c^2} = \sqrt{1 - u^2/c^2}\, l'$$

As measured in K the length of the rod will be *less* than as measured in K' by a factor $\sqrt{1 - u^2/c^2}$, which depends on the velocity of the rod with respect to K. *Lengths depend in a particular way upon velocity* in relativistic kinematics. Length (or spatial interval), like velocity, is a magnitude which is *relative to the frame of reference with respect to which it is measured.*

A similar result obtains for temporal intervals (durations). Consider a clock at rest with respect to K' and a temporal interval τ' between two ticks of the clock at t_1' and t_2' (so that $\tau' = t_2' - t_1'$). What is the temporal interval between the two ticks as measured in K, that is, what is $\tau = t_2 - t_1$?

Solving the first and fourth Lorentz equations for the unprimed coordinates, we get

$$x = \sqrt{1 - u^2/c^2}\, x' + ut \qquad \ldots (1)$$

$$t = \sqrt{1 - u^2/c^2}\, t' + \frac{ux}{c^2} \qquad \ldots (2)$$

From eq. (2) we get

$$\tau = t_2 - t_1 = \sqrt{1 - u^2/c^2}\,(t_2' - t_1')$$

$$+ \frac{u}{c^2}(x_2 - x_1) \ \ldots (3)$$

But from eq. (1)

$$x_2 = \sqrt{1 - u^2/c^2}\, x_2' + ut_2$$

and

$$x_1 = \sqrt{1 - u^2/c^2}\, x_1' + ut_1$$

Now since the clock is at rest with respect to K', $x_2' = x_1'$, so that

$$x_2 - x_1 = u\,(t_2 - t_1)$$

Putting this result into eq. (3), we obtain

$$\tau = t_2 - t_1 = \sqrt{1 - u^2/c^2}\,(t_2' - t_1') + \frac{u^2}{c^2}(t_2 - t_1)$$

Solving for $t_2 - t_1$, we get

$$t_2 - t_1 = \frac{1}{\sqrt{1 - u^2/c^2}}(t_2' - t_1')$$

or

$$\tau = \frac{1}{\sqrt{1 - u^2/c^2}}\, \tau'$$

As measured in K the temporal interval between the two ticks of the clock is *greater* than as measured in K' by a factor $1/\sqrt{1-u^2/c^2}$, which depends on the velocity of the clock with respect to K. *Temporal intervals, or dura-*

tions, depend in a particular way upon velocity in relativistic kinematics. Duration, too, is a magnitude which is *relative to the frame of reference with respect to which it is measured.*

These consequences of *LT* have usually been expressed by saying that moving rods *shrink* and moving clocks *slow down.* This manner of stating the situation can be misleading, for it may suggest that the "shrinkage" and "slowing" are *effects* of velocity and thus somehow due to velocity-dependent *forces.* When Lorentz originally introduced *LT*, he did so in just this way, as expressing the *effects* on bodies of *electromagnetic forces* among the ultimate constituents of matter: as velocity-dependent forces (velocity with respect to the ether) which *cause* bulk matter to shrink in the direction of its motion through the ether. According to the present point of view, however, we are dealing, *not* with *dynamical effects*, but with the kinematical background of any dynamical theory.[4] In other words, we are considering *LT* and the consequences of *LT* as expressing the very *structure of spatial and temporal relations.* The relativity of length and duration to a frame of reference does not therefore depend upon any particular theory about the ultimate structure of matter. This is one of the most revolutionary aspects of that general body of knowledge known as the "theory of relativity."

It is *relativistic kinematics* which has engaged most of the attention of philosophers who have written on relativity theory. They have not paid as much attention to the new, or relativistic, *dynamics*, to which I shall now turn.

---

[4] Lorentz was not as successful in explaining dynamically the "slowing" of moving clocks as he was with respect to the "shrinkage" of moving bodies. His theory thus fails on less abstract grounds as well as on grounds depending on the basic distinction between kinematics and dynamics.

THE NEED FOR A NEW DYNAMICS

We saw that the basic laws of classical *dynamics*, including Newton's second law of motion $\mathbf{F} = m\mathbf{a}$, are not invariant under LT, and there thus seems to be a unique "mechanical frame" in which $\mathbf{F} = m\mathbf{a}$ and the various specific force laws do hold. In other words, *classical dynamics is not a relativistic theory* (except for one branch, electrodynamics).

There are two courses of action that can be taken in the face of these startling results (the strange new kinematics and the noninvariance of $\mathbf{F} = m\mathbf{a}$ under LT). We could be "conservative," taking the present situation as an indication that we have gone badly astray, and return to the old kinematics of GT and go back to the Galilean Principle of Relativity. This would entail attempting once again to construct Lorentz-type theories and/or modifying electrodynamics.

On the other hand, we can stick to the new kinematics of LT, assume that the Special Principle of Relativity *is* correct with LT, and conclude:

Since $\mathbf{F} = m\mathbf{a}$ *is not invariant under LT, it is not a correct law of nature after all! We must therefore construct a new theory of dynamics.*

This latter alternative produces the following rather ironic situation: The belief in a relativity principle of some sort—born of classical dynamics and refined and extrapolated on the basis of one branch of classical dynamics, electrodynamics—comes back to destroy the basic laws of classical mechanics (both kinematics and dynamics)!

It is this latter revolutionary alternative that was chosen by Einstein. We must seek a new foundation for dynamics, a new general force law to replace $\mathbf{F} = m\mathbf{a}$. The dynamical theory based upon the new foundation must conform to the following conditions:

(1) Its laws must be invariant when transformed among

frames of reference moving at constant velocity with respect to each other—under a *Lorentz* transformation. (This is one form of the Special, or "Restricted," Principle of Relativity. See Chapter 7.)

(2) Its laws must reduce to those of *classical*, or Newtonian, dynamics for those ranges of the relevant variables where classical dynamics is well-confirmed (in particular, when $u^2/c^2 << 1$). That is, the classical theory must be a formally limiting case of the new *relativistic theory*.

It should be noted that these are very general and sweeping requirements, not only requirements for the new dynamics. The Special Principle of Relativity requires *every* physical theory to be "Lorentz-invariant" (condition (1)). It demands that we review all the "old" theories we have and modify them, if necessary, to make them Lorentz invariant—to make them "relativistic." The Special Principle of Relativity therefore provides a directive to be used in *all* branches of physics: When formulating laws in a given domain, try explicitly to make them Lorentz-invariant.[5]

There is a third condition which the new dynamics must also meet: When the consequences of the new dynamics disagree with those of the old, the new dynamics must of course be correct. But this is just the requirement of agreement with observation that is demanded of any acceptable theory. Ultimately, therefore, it is an empirical question as to which of the two alternatives mentioned above is to be accepted. This may seem too obvious to be worthy of mention, but it is useful to be reminded of this point after the argumentation of the last few chapters, some of which might otherwise appear to be armchair speculation.

In the following chapter I shall present the new, *relativistic dynamics*.

[5] Previously, physicists implicitly tried to formulate laws which were "Galileo-invariant."

10

RELATIVISTIC DYNAMICS

How shall we go about finding a replacement for Newton's second law of motion? We begin by recalling, from the opening paragraphs of Chapter 2, that Newton's second law can be written as

$$\mathbf{F} = \frac{d}{dt}(m\mathbf{v}) \qquad \ldots (1)$$

(This will reduce to the form $\mathbf{F} = m\mathbf{a}$ if the mass m does not vary with time.) Let us introduce the symbol 'p' as shorthand for the expression '$m\mathbf{v}$', and refer to this product as the 'momentum' of the body in question. Eq. (1) can then be rewritten as

$$\mathbf{F} = \frac{d\mathbf{p}}{dt} \qquad \ldots (2)$$

We shall now look upon eq. (2) as being "more basic" than eq. (1). By this I mean that we shall consider Newton's second law of motion as stating that the force acting upon a body is equal to the first derivative with respect to time of the *momentum* of the body, where momentum is now to be understood as *some as-yet-to-be-determined function* of the body's mass and velocity, m and $\mathbf{v}$. The product $m\mathbf{v}$ would then be *one possibility* for this function. That is, from this perspective eq. (1) is but *one possible instantiation* of eq. (2). We have used eq. (1) as a ladder to reach eq. (2), and then we throw away the ladder. There are now a number of possibilities of descent, only *one* of which is letting p be equal to *m*v. In fact, this particular possibility is ruled out, for we saw in the previous chapter that the statement expressed by eq. (1)—Newton's second law of motion—is not invariant under a Lorentz transformation between frames of reference moving uniformly with respect to each other!

Our original problem is thus transformed into the following problem: Find a *new* function p of *m* and v such that eq. (2) is acceptable, where 'acceptable' means that $\mathbf{F} = d\mathbf{p}/dt$, with the proper p, satisfies the following conditions:

a) It is Lorentz-invariant;
b) It reduces to $\mathbf{F} = m\mathbf{a}$ for low velocities;
c) It leads to correct predictions.

These are just the three conditions discussed toward the end of the last chapter.

What function of *m* and v shall we use for p? If we use

$$\mathbf{p} = \frac{m\mathbf{v}}{\sqrt{1 - v^2/c^2}} \qquad \ldots (3)$$

then the force law

$$\mathbf{F} = \frac{d}{dt}\left(\frac{m\mathbf{v}}{\sqrt{1 - v^2/c^2}}\right) \qquad \ldots (4)$$

will satisfy conditions a) – c) and is therefore the appropriate replacement for $\mathbf{F} = m\mathbf{a}$, and will serve as the foundation of the new, *relativistic dynamics*.

There are a number of ways of arriving at the conclusion that eq. (3) expresses the appropriate functional form for p. For example, one could simply "guess" it. The guess is right if and only if the result satisfies conditions a) – c). This "procedure" is more respectable than it may appear, since, to an experienced physicist, putting in the $\sqrt{1-v^2/c^2}$ term stands out as a likely candidate, given the form of the Lorentz transformation equations. (See Chapter 8, p. 106.)

Another procedure is the following: In classical dynamics there is a theorem that can be derived from Newton's laws of motion which states that *momentum is conserved in a closed system*. That is, if we have a system of particles in motion, exerting forces on one another, and such that there are no forces exerted on the particles from outside the system, then the total momentum (defined as $\mathbf{p} = m\mathbf{v}$) of all the particles in the system will not change with time. This law is invariant under a *Galilean* transformation. But it is *not* invariant under a *Lorentz* transformation. That is, if the sum of the momenta of the particles is constant when measured in a frame K, we would expect it to *vary* with time in another frame K' moving uniformly with respect to K *if LT is used* to effect the transformation of the appropriate parameters.

If we now *require* that there be a momentum conservation theorem in the new dynamics, where momentum is *some new function* of m and $\mathbf{v}$, and also require that this theorem be *Lorentz*-invariant, we can *deduce* that $\mathbf{p}$ must be as given by eq. (3).[1]

This way of obtaining a replacement for Newton's second law of motion is an instance of a general type of strategy used in theoretical physics which philosophers of science seem not to have noticed.

The general procedure is this: There is a law which has worked well in the past, but new considerations (either

[1] For the details of the derivation see Max Born, *Einstein's Theory of Relativity* (Dover, 1962), pp. 267 ff. (The first edition of this book appeared in 1920.)

theoretical or experimental) indicate that the law is incorrect after all. We do not, however, simply discard the old law and start from scratch, for we wish to trade on the fact that the old law was in some sense on the right track. Instead, we *rewrite the old law in a more abstract form,* and then apply the new considerations in order to reach the new law. Both the old and the new laws will then appear as (different) instances of the abstract form, but the new one will be correct because it conforms to the conditions elicited from the new considerations.

In the present case the abstract form of the old law, $\mathbf{F} = d/dt(m\mathbf{v})$, is $\mathbf{F} = d/dt$ (some function of m and $\mathbf{v}$). The new considerations which are used to obtain the new function of m and $\mathbf{v}$ are two in number. First, the new function must be such that the resulting new law is Lorentz-invariant (rather than Galileo-invariant). Second, it is required that the new function lead to a Lorentz-invariant conservation law (just as the old function, $m\mathbf{v}$, led to a Galileo-invariant conservation law).

The original law then appears as but one "version" of the abstract form, in which (from the perspective of the new law) certain variables have been allowed to take certain limiting values.

(There are some applications of this general strategy in which the abstract form of the old law may be allowed to contain variables which do not even appear in the old law. In this way *quantum mechanics* can also be "reached" from classical mechanics. In the present case no new variables were introduced; we only needed a new "arrangement" of the old variables.)

The result of this type of argument is the appearance of the original law as a "good approximation" for certain ranges of the relevant variables (including any new variables which may have been introduced in the process of abstraction from the original law).[2]

We have then in eq. (4) our new "second law of motion" —the new law which states in general what the effect is of a force acting upon a massive body. It was constructed in such a way as to conform with conditions a) and b) at the beginning of this chapter, and it has been shown experimentally in the years since its introduction that it also satisfies condition c).

RELATIVISTIC DYNAMICS

Similar procedures can be used to generate replacements for the *specific* force laws for specific situations in which massive bodies find themselves. In this manner a new theory of dynamics can be developed: *relativistic dynamics.*

One of the advantages of having spent so much time treating *classical* mechanics now becomes apparent. For the general function and structure of the new theory of mechan-

---

[2] The application of the general strategy to the case at hand has an interesting twist. In classical dynamics momentum ($= mv$) conservation is, logically, simply one theorem among many which are deducible from $\mathbf{F} = m\mathbf{a}$. Therefore one might expect that when we discard $\mathbf{F} = m\mathbf{a}$, the conservation of momentum theorem would also be discarded. But conservation of momentum is such an important theorem that we required the replacement for $\mathbf{F} = m\mathbf{a}$ to also yield a conservation of momentum theorem. (Of course, in the new, relativistic dynamics the quantity which is labeled 'momentum' must therefore be different from that of classical dynamics.)

This logical point has an interesting historical antecedent. Leibniz, in his physics, was able to discover the conservation of momentum without being able to deduce it from his dynamical theory. That is, he could argue for this theorem without invoking any general force law.

ics can now be stated quite briefly. We can simply paraphrase most of what was said in Chapters 2 and 3! That is,
relativistic dynamics can be presented in quasi-axiomatic
form just as *classical* dynamics was presented in earlier
chapters, now that we have developed the *foundations* of
the theory as a response to difficulties in classical dynamics
(including electrodynamics) in the chapters preceeding this
one.[3]

Paraphrasing the opening paragraphs of Chapter 2, for
example, we obtain

Relativistic mechanics—more specifically, relativistic dynamics—deals with the kinds of motions a system of massive bodies will undergo when subjected to the influence of
different kinds of forces.

We postulate the following law of nature as stating the
effect of a force acting on a body:

When a force **F** acts upon a body of mass m and velocity
v at a time t, the body will respond in accordance with the
equation

$$\mathbf{F}(t) = \frac{d}{dt}\left(\frac{mv(t)}{\sqrt{1 - v^2/c^2}}\right)$$

We could then speak again of various specific force laws
for different kinds of situations in which a massive body
might find itself, distinguish again between the two types of
mechanical problem (find the motion when the force law is
given; find the force law responsible for a given motion),
and so on. Of course, the specific force laws would be, not

---

[3] Notice that we could have presented classical dynamics as a
response to difficulties in older views, rather than in the quasi-
axiomatic manner of Chapters 2 and 3. (It should be obvious that
axiomatic and historical approaches are equally "correct"; they
simply serve different purposes and highlight different aspects of
any theory under consideration.)

those introduced in the early chapters, but rather Lorentz-invariant laws which formally reduce to the old laws when $v \ll c$. But there would be the same or similar conventions for determining what constitutes a mechanical explanation (rather than an engineer's formula), the same or similar constraints, and so on.

The relativistic mechanics of rigid bodies, elastic bodies, and fluids could be introduced as they were before, and we could discuss more abstract formulations of the theory: analytical relativistic mechanics.[4] (Of course, that branch of classical dynamics known as *electrodynamics* will emerge unscathed. *Relativistic* electrodynamics = *classical* electrodynamics, for the equations of the classical theory were already Lorentz-invariant!) What I am saying can be summed up by pointing out that in the label 'relativistic mechanics', the word 'mechanics' should not be lost sight of. *We still have a theory of the effects of forces on the motions of bodies* whose general structure and function is the same as that of classical mechanics. *In this respect* it is not conceptually different from the theory which it replaces. (It is of course *relativistic* (or Einsteinian) mechanics rather than *classical* (or Newtonian) mechanics, with all the mathematical changes which this difference of terminology encapsulates.)

"RELATIVITY THEORY"

It will be useful to conclude this chapter by clarifying a few bits of terminology which are not always used very

---

[4] See the last page of Chapter 2.

There will also be some interesting theoretical differences. For example, one consequence of relativistic mechanics is that there cannot exist a perfectly rigid body! Another such difference, regarding the relationship between mass and energy, will be discussed in the next chapter.

clearly in the more popular literature on relativity theory. (Many physics texts are also unclear in this regard.) I shall do so by laying down some suggested definitions.

Restricted (or Special) Principle of Relativity

Form (1): Laws of nature should be invariant when transformed between frames of reference moving uniformly with respect to each other. (In this form, with no mention of which transformation equations are to be used, Newton could accept the principle, tacitly using the Galilean transformation equations.)

Form (2): Laws of nature should be invariant under a *Lorentz* transformation between frames moving uniformly with respect to each other. (In this form the principle *is* new and serves as a general directive in areas of physics other than mechanics.)

Form (1), together with the law of the constancy of the velocity of light, yield the Lorentz transformation equations, and thus form (2) of the principle.

Relativistic Mechanics

This is the theory of motion which conforms to form (2) of the Restricted Principle of Relativity. It is one particular physical theory, including both a kinematics and a dynamics.

Special Theory of Relativity

This is the catchall term for everything we have been doing in the last few chapters, that is, that general body of knowledge containing both the Restricted Principle of Relativity and relativistic mechanics. (The *general* theory of relativity would then be that general body of knowledge including a *General* Principle of Relativity, in which *arbitrary* relative motions of reference frames are considered.)

So *relativistic mechanics* is simply the new correct physical theory of force and motion, whereas *classical mechanics* is the older and incorrect physical theory for the *same general physical domain.*

There are some consequences of relativistic mechanics which are importantly different from those of classical mechanics, and I shall discuss some of them in the following chapter. They will be recognized as some of the startling results that popularizations of relativity theory fasten upon as central, thereby giving a distorted view of the nature of the theory. I shall attempt to keep these consequences in their proper perspective.

11

THE MASS-ENERGY RELATION

I said in the last chapter that classical and relativistic dynamics are not conceptually different in the sense that they are both theories which deal with the effects of forces on the motions of massive bodies. In fact, the equations of classical dynamics are formally deducible from those of relativistic dynamics in the limit of low velocities. This might seem to suggest that the two theories differ only quantitatively for velocities which are not small in comparison with the velocity of light. But such a conclusion does not follow automatically, for there are some theorems in relativistic dynamics which are surprising enough to have led some to conclude that the two theories are qualitatively different. These theorems have given rise to a discussion of whether one can, after all, look upon classical dynamics as "merely" a certain limiting case contained within the relativistic theory.

In particular I have in mind the famous equation $E = mc^2$ and the notion that according to relativity theory matter and energy are "interconvertible"—an equation and an idea which everyone who reads his newspaper is at least aware of. It has recently been argued that results such as this show that the *apparent* deducibility of classical dynamics from relativistic dynamics for the limit of low velocities is spurious—at best a formal trick—and that the classical and relativistic theories are somehow *conceptually incommensurable* since certain key terms *used* in both theories, such as 'mass' and 'energy', have *different meanings* in the two theories.[1] The supposed meaning change is allegedly due to the appearance in the relativistic theory of theorems such as $E = mc^2$, which do not occur in the classical theory. It is then further argued that these alleged changes in meaning of key terms are "central to the revolutionary impact of Einstein's theory" and result in "a displacement of the conceptual network through which scientists view the world."[2]

In my opinion the validity of these and certain related claims cannot be responsibly assessed without a clear understanding of the precise way in which concepts such as *energy* are introduced and used in both classical and relativistic dynamics, and without seeing exactly how the equation $E = mc^2$, for example, is derived within the relativistic theory. It is only on the basis of a methodologically clear presentation of the physics involved that one can convincingly argue about the question of meaning change and related philosophic issues.[3]

In this chapter I shall not attempt to resolve fully the issue of meaning change with respect to the key concepts of classical and relativistic dynamics, although I will have a few remarks to make on this issue. I will instead be more concerned with an attempt to delineate clearly the way in which the "spectacular" theorem $E = mc^2$ is derived and discuss the role which it plays in relativistic dynamics. If I

[1] See T. S. Kuhn, *The Structure of Scientific Revolutions* (University of Chicago Press, 1962), Chapter 9.

[2] *Ibid.*, p. 101. Kuhn goes on to speak of scientific revolutions— the transition from classical to relativistic dynamics being a prototype—as "changes of world view." See Chapter 10 of his book.

[3] At the risk of being accused of moralizing, I would make the general claim that many inconclusive discussions in philosophy of science would come closer to resolution if philosophers paid closer attention to the physics (psychology, biology, and so on) involved.

am successful, this will be a large step toward resolution of the philosophic issues which seem to center on this equation.

ENERGY IN CLASSICAL DYNAMICS

I shall begin by introducing some notions in classical dynamics which I did not discuss in earlier chapters.

In the classical theory it is of interest to consider the cumulative or integrated effect of a force acting on a body. This can be done in two ways. One may think of the cumulative effect over the *time* interval during which the force acts or over the *spatial* distance traversed by the body while the force acts. The latter is called the *work*, *W*, done by the force on the body and is defined by the equation

$$W = \int_{\mathbf{r}_1}^{\mathbf{r}_2} \mathbf{F} \cdot d\mathbf{r} \qquad \dots (1)$$

where $\mathbf{r}_1$ and $\mathbf{r}_2$ are the initial and final positions of the body.[4] If the motion of the body is in a straight line (along the X-axis, say) and the force acts along this line,[5] equation (1) becomes

$$W = \int_{x_1}^{x_2} |\mathbf{F}| \, dx \qquad \dots (2)$$

If we now substitute $m(d\mathbf{v}/dt)$ for $\mathbf{F}$ in accordance with

[4] The force $\mathbf{F}$ is a vector, and the symbol '$d\mathbf{r}$' denotes a "small" vector tangent to the trajectory of the particle. The "dot product" of these two vectors, $\mathbf{F} \cdot d\mathbf{r}$, is defined as the magnitude of $\mathbf{F}$ multiplied by the magnitude of $d\mathbf{r}$, multiplied by the cosine of the angle between these two vectors. The work W is thus a *scalar* quantity.

[5] The results which will be of importance for our purposes follow in this simpler case as well as the more general case in which the body traverses some curvilinear path and the force acts at some (possibly changing) angle to its path. (If the force is constant in magnitude, equation (2) becomes simply $W = F(x_2 - x_1)$.)

Newton's second law of motion, eq. (2) can be rewritten as

$$W = \int_{x_1}^{x_2} m \frac{d\mathbf{v}}{dt} dx$$

But $\mathbf{v} = dx/dt$, or $dx = vdt$, so the last equation becomes

$$W = \int_{v_1}^{v_2} m v \, dv$$

where 'v_1' and 'v_2' represent the velocities of the body at its initial and final positions. Carrying out the integration, we obtain

$$W = \frac{1}{2} mv_2{}^2 - \frac{1}{2} mv_1{}^2 \qquad \ldots (3)$$

Again, for simplicity, let us assume that the body was initially at rest. That is, $v_1 = 0$. We then get

$$W = \frac{1}{2} mv_2{}^2 \qquad \ldots (4)$$

That is, the cumulative effect of a force acting on a body, when the effect is summed (integrated) over the *distance*, is an increase in the quantity $\frac{1}{2} mv^2$ for the body.

Why should this result be of any particular interest? After all, could we not have arbitrarily defined a quantity **B**, say, by the equation

$$\mathbf{B} = \int_{r_1}^{r_2} \mathbf{F} \times d\mathbf{r}^6$$

and carried out the integration to see what happens? (We could call this quantity the 'bork' done by the force on the body.)

The *work* is of interest because the particular quantity

---

[6] $\mathbf{F} \times d\mathbf{r}$ is called the "cross-product" of the two vectors. It is a *vector*, not a scalar.

$\frac{1}{2}mv^2$ (as opposed to, say, $\frac{2}{3}\sqrt{mv^4}$) *enters into a certain conservation law* that can be developed within classical dynamics. That is, for certain kinds of simple systems one can deduce that whenever the quantity $\frac{1}{2}mv^2$ *increases* (decreases) for one body of the system, the quantity $\frac{1}{2}mv^2$ summed over the other bodies of the system *decreases* (increases) by a compensating amount. (We say that the other bodies have "done work" on the body in question.) It is therefore useful to give the quantity $\frac{1}{2}mv^2$ a name— *energy* (actually *kinetic energy*)[7]—in order to focus attention upon it.[8]

There sometimes arises the temptation to substantialize the concept of energy, to treat it as if it referred to some sort of "stuff" which is conserved. This would be a mistake. At best, energy would be a rather strange sort of "stuff," portions of which could go out of existence at one place while other portions come into existence elsewhere with no intervening activity, no *continuous transference* of "it" through space and time. (This is the way in which one would be forced to describe the activity occurring in certain dynam-

---

[7] I have oversimplified the situation a bit. The conservation law I have in mind involves not only the quantity $\frac{1}{2}mv^2$ but also a certain other quantity—called 'potential energy'—whose precise functional form depends on the force laws for the particular systems being considered. That is, in the "certain kinds of simple systems" which I have in mind, it is the sum of the *kinetic energies* ($\frac{1}{2}mv^2$) and the so-called *potential energies* which is conserved (where 'conserved' means 'does not vary with time').

(There are also some "extremely simple" systems where kinetic energy itself is conserved. These are systems in which bodies exert forces on one another only when they are in contact.)

[8] The concept of *momentum* and its related conservation law (briefly discussed in Chapter 10) are introduced in a manner similar to that just used for energy, but the cumulative, or integrated, effect of a force acting on a body is calculated over the *time* interval during which the force acts.

ical interactions if he insisted on treating energy as quasi-material.)

Now there are certain kinds of systems in which *energy* (kinetic *plus* potential; see footnote 7) is *not* conserved. This would seem to imply that there was no point in paying special attention to the quantity $\frac{1}{2}mv^2$ in the first place, or that at best the energy concept has only a limited usefulness for a subset of mechanical systems. (In other words, it appears that *work* is not much more interesting or less arbitrary than *bork*.)

However, it turns out to be possible to find other parameters of such systems: *nonmechanical parameters*[9] (such as the temperatures of the bodies of the system) such that when the energy (kinetic plus potential) has decreased (increased), certain functions of these nonmechanical parameters have increased (decreased) by a compensating amount. That is, while energy (in the sense of $\frac{1}{2}mv^2$ plus potential energy) is not conserved in certain systems, energy plus certain functions of other (nonmechanical) parameters of such systems *is* found to be conserved.

This *fact* leads physicists to refer to $\frac{1}{2}mv^2$ plus potential energy as *mechanical* energy and to invent similar terms of qualification for the appropriate functions of the nonmechanical parameters, thus referring to them as *other forms of energy*. In this way we have a *law of conservation of energy which transcends classical dynamics*.

It must be emphasized that as far as classical dynamics is concerned this is a bit of "cheating" that the experimental results of *other theories* fortunately make possible. It is an (fortunate) *empirical fact* that when "mechanical energy"

---

[9] A nonmechanical parameter is one which appears in no force laws. It does not contribute to the forces acting upon a body. See Chapter 3, p. 32.

is not conserved in certain systems, other "forms of energy" can be *constructed* within the descriptive apparatus of other theories to preserve the "law" of conservation of (the generalized notion of) "energy." Nature was kind to us in this regard.

As far as classical dynamics is concerned, this law of conservation of energy (in the generalized sense of 'energy') is left unexplained. It is not a logical consequence of the laws of dynamics ($F = ma$ plus specific force laws)! It may be viewed as an inductive generalization "derived" from the successful attempts to construct "other forms of energy" for certain kinds of systems, as described above.[10]

Let us briefly look at a simple example of a situation in which "mechanical energy" is not conserved and in which "other forms of energy" can be constructed to preserve the general law of conservation of energy. Consider two equal masses of putty or clay moving toward one another in a straight line with equal but oppositely directed velocities v_A and v_B. The total mechanical energy of the two bodies of the system is

$$m|v_A|^2$$

since the energy of each is $\frac{1}{2}m|v_A|^2$ (given that their

---

[10] The law of conservation of energy has in fact assumed the role of an *a priori* truth in some contexts. It was the insistence on its truth in the face of an apparent exception that eventually led to the discovery of the neutrino, a particle originally "invented" to explain the apparent loss of energy in certain kinds of nuclear reactions.

Notice in this example how the concept of energy, originally introduced in classical dynamics, comes to play a role in a domain—elementary particle physics—where classical dynamics is hopelessly inadequate. Just as in the case of momentum conservation discussed in Chapter 10, we have a theorem within a physical theory which is "saved" even though the original theory is abandoned. (In this case, too, the key concept of the theorem—energy—must assume a different functional form.)

masses are both equal to *m* and that $|\mathbf{v}_A| = |\mathbf{v}_B|$).[11] When
the bodies collide, they will coalesce into one body with
zero velocity.[12] The total mechanical energy of the system
after the collision is zero. The mechanical energy of the
system has decreased; it has not been conserved.

But it is observed in situations such as this that the *temperature* of the resulting blob of clay is higher than the
temperature of the two bodies before impact (assuming that
their temperatures were equal to one another before impact).
In fact, one can construct a function of the temperature and
other *thermodynamic parameters* of this system such that
the difference in the measured values of this function before
and after impact turns out to be equal in magnitude to the
value of the "lost" mechanical energy. In this way one can
introduce the concept of *heat energy*. (The damped spring
discussed in Chapter 2 is another such system.)

Now the general Newtonian program discussed in Chapter 3 dictates that in situations such as that just described
we try to reduce (in the sense of Chapters 3 and 4) the
parameters of the *other theories* (thermodynamics, in the
example) to those of mechanics. In other words, we should
try to show that the "other forms of energy" are at bottom
mechanical in nature. In fact, nineteenth-century physics was
quite successful in reducing *heat energy* to *mechanical energy*: the kinetic energy of the atoms of which matter (the
blobs of clay, for example) is composed.

It should be clear that there is nothing "wrong" with a

---

[11] I am assuming that there are no noncontact forces which would
give rise to a potential energy term, so that the total mechanical
energy is simply the sum of the kinetic energies of the bodies. (One
can picture the bodies moving toward one another in empty space,
far removed from all other bodies.)

[12] This is referred to as a "totally inelastic collision."

form of energy which cannot be so reduced. Lack of success in this regard merely shows that there are more "things" in this universe than can be handled with the descriptive apparatus of classical mechanics. And I trust that this comes as a surprise to no one.

There is more that could be said about the concept of energy in classical physics, but the preceding is sufficient as a background against which we can examine the manner in which the concept functions in relativistic dynamics.

ENERGY IN RELATIVISTIC DYNAMICS

Quite generally, when elaborating a new theory for a particular domain in which an older theory is in some sense a "good approximation" for certain ranges of the relevant parameters, it is a useful procedure to attempt to parallel as much as possible the manner in which the older theory had been elaborated. In developing relativistic dynamics, then, we should try to do the same sorts of things which we did in the classical theory, but by using the new force law (and, of course, the Lorentz transformation equations whenever it becomes necessary or convenient to change reference frames).

So, in relativistic dynamics we once again introduce the concept of work, in the *same manner* as in classical dynamics: as the integrated effect (calculated over distance) of a force acting on a body. That is,

$$W = \int_{r_1}^{r_2} \mathbf{F} \cdot d\mathbf{r} \qquad \dots (1) \; again!$$

Considering once again the one-dimensional case for simplicity, this becomes

$$W = \int_{x_1}^{x_2} |\mathbf{F}| \, dx \qquad \dots (2)$$

But now we substitute for **F** in accordance with the *relativistic force law* (Chapter 10, p. 124, eq. (4)) to obtain

$$W = \int_{x_1}^{x_2} \frac{d}{dt}\left(\frac{mv}{\sqrt{1-v^2/c^2}}\right) dx \qquad \ldots \text{(5)}$$

As before, $dx = vdt$, so eq. (5) can be rewritten as

$$W = \int_{v_1}^{v_2} v\, d\left(\frac{mv}{\sqrt{1-v^2/c^2}}\right)$$

Carrying out the integration, we get

$$W = \frac{mc^2}{\sqrt{1-v_2^2/c^2}} - \frac{mc^2}{\sqrt{1-v_1^2/c^2}} \qquad \ldots \text{(6)}$$

This is the relativistic analogue of eq. (3). We can get the relativistic analogue of eq. (4) if we assume again (for simplicity) that the body was initially at rest: that $v_1 = 0$. Eq. (6) then becomes

$$W = \frac{mc^2}{\sqrt{1-v^2/c^2}} - mc^2 \qquad \ldots \text{(7)}$$

This expression[13] for the work W looks radically different from the classical expression given by eq. (4). But let us expand eq. (7) in powers of v/c. We get[14]

---

[13] I have dropped the subscript '2' on the 'v'.

[14] In general, whenever $x^2 < 1$, the series

$$1 + \frac{1}{2}x + \frac{1\cdot 3}{2\cdot 4}x^2 + \frac{1\cdot 3\cdot 5}{2\cdot 4\cdot 6}x^3 + \ldots$$

approaches the limit

$$\frac{1}{\sqrt{1-x}}$$

In our case 'x' stands for v^2/c^2.

$$W = mc^2 \left(1 + \frac{1}{2}\frac{v^2}{c^2} + \frac{3}{8}\frac{v^4}{c^4} + \frac{5}{16}\frac{v^6}{c^6} + \dots \right) - mc^2$$

or

$$W = \frac{1}{2}mv^2 + \frac{3}{8}\frac{m}{c^2}v^4 + \frac{5}{16}\frac{m}{c^4}v^6 + \dots \quad \dots \quad (8)$$

(The mc^2 terms conveniently cancel each other.) Now if v is "much smaller" than c ($v << c$), the higher terms of the series become "very small," so that eq. (8) can be approximated by its first term, yielding

$$W \simeq \frac{1}{2}mv^2$$

But this is just the classical expression for the work done by the force on the body! So, in spite of the rather different expressions for the work W in classical and relativistic dynamics (equations (4) and (7)), we see that the relativistic expression reduces to the classical expression for low velocities, as required. (See Chapter 9, p. 121.)

Let us now go back to the exact expression for the work given by eq. (7), but with the symbol 'm_0' in place of 'm' (for reasons which will be given below):

$$W = \frac{m_0 c^2}{\sqrt{1 - v^2/c^2}} - m_0 c^2 \quad \dots \quad (9)$$

Interpreting this in the same manner as in the classical case, we must say that the integrated effect of the force is to increase the *energy* of the body. So the first term in eq. (9)

$$\frac{m_0 c^2}{\sqrt{1 - v^2/c^2}}$$

represents the total (mechanical) energy of the body after the force has acted, and the second term, $m_0 c^2$, represents the energy of the body when the force began to act (so that the difference is the *increase* in energy due to the force).

Now in the classical case the second term was equal to zero, since we had assumed the body to be initially at rest. (See the transition from eq. (3) to eq. (4), where $\frac{1}{2}mv_1^2 = 0$ since $v_1 = 0$.) But in the relativistic case the second term did *not* become equal to zero, even though we assumed again that $v_1 = 0$, because of the different functional form for the relativistic expression for W. (See the transition from eq. (6) to eq. (7).)

In other words, it is a consequence of relativistic dynamics that *a body at rest has a nonzero mechanical energy*[15] *given by the product of its mass and the square of the velocity of light:* m_0c^2. This has been called the "rest energy" of the body.[16]

It should be observed that m_0c^2 will usually be a rather large number. For the normal sorts of situations dealt with in classical physics, where the velocities involved are small in comparison with c, the quantity m_0c^2 will in fact be larger by many orders of magnitude than the kinetic, thermal, and other "forms of energy" involved.[17] It might therefore be asked why—if relativistic dynamics is correct—

[15] This is totally apart from any (additional) potential energy which the body might have.

[16] This result has nothing to do with any changes of reference frame. It has been deduced with (tacit) reference to *one* frame of reference within which the relevant parameters might be measured.

To be consistent with expressions such as 'thermal energy' (energy due to thermal properties) and 'kinetic energy' (energy due to motion), it would seem more appropriate to use the appellation 'mass energy' (energy due to mass) for the quantity m_0c^2. It sounds odd, after all, to say "energy due to rest" (for 'rest energy'). But I shall follow the customary usage in what follows.

[17] The numerical value of c is 3×10^{10} cm/sec, so that $c^2 = 9 \times 10^{20}$ cm^2/sec^2.

the rest energy had not previously been observed to be of importance. Why was its existence not "inductively" inferred from the experimental results of classical nineteenth-century physics, before its existence was deductively inferred within relativistic dynamics?

The reason is that it is *changes* of energy which are of importance. The importance of the concept of energy lies in its *conservation law*, the statement that when one form of energy in a closed system *changes* (increases or decreases) other forms will change (decrease or increase) by a compensating amount. Therefore, as long as the masses (m_0) of the bodies in a system remain *constant*, their rest energies (m_0c^2) will remain constant and will, as it were, go unnoticed. That is, there would be no need to take them into account when using the law of conservation of energy to describe, explain, or predict what takes place in a physical system. (Similarly, if the temperatures of bodies always remained constant, the concept of thermal energy would probably not have been developed.)

One could almost define the concept of a *form of energy* in a quasi-recursive manner as any function of any set of parameters of a closed physical system such that it is required to preserve the truth of the law of conservation of kinetic energy (introduced through the work concept in mechanics) *plus* mechanical potential energy *plus* any other such functions *previously* found necessary to preserve . . . (and so on). . . . Thus, there was never any occasion in nineteenth-century physics necessitating the introduction of *rest energy* in order to preserve . . ., since no *changes* in what we now call 'rest energy' were ever observed.

The relativistic situation is the same. That is, *if* the masses (m_0) of bodies remain constant, we would never have occasion to make use of the concept of the rest energy (m_0c^2) of bodies. Rest energy would be, as it were, a form of energy "locked into" bodies by virtue of their constant mass, which could never be used to do work on other bodies.[18]

MASS AND ENERGY IN RELATIVISTIC DYNAMICS

The remarks of the last few paragraphs rested on the assumption that the masses of bodies remain constant.[19] This assumption had been erected into a *principle of the conservation of mass*: The total mass of the bodies in a closed system does not vary with time. This principle or law has a different logical status than that of the conservation of energy. It is wholly disconnected with the rest of classical physics in the sense that it cannot be deduced from the laws of classical dynamics at all. Energy conservation arises out of an extrapolation from certain constructs in classical dynamics, whereas mass conservation is imposed "from the outside" as a kind of boundary condition on mechanical problems.

Let us ask what the consequences would be if we dropped the assumption of the constancy of mass. Suppose that the mass m_0 of some body in a closed system became zero.[20]

---

[18] Those with positivistic inclinations might then infer from this that the notion of rest energy, read out of the nonvanishing second term of eq. (9), is a kind of meaningless oddity within relativistic dynamics which mars an otherwise highly useful and elegant theory.

[19] There is a trivial sense in which this assumption is almost always false. For example, my pen decreases in mass as ink flows out of it and increases in mass when I fill it. Most of the objects on my desk slowly decrease in mass through time with wear and tear. But, obviously, within classical physics cases like this are interpreted as the *transference* of smaller bodies, each of which has a constant mass.

[20] I am supposing that this is not due to "transference," as discussed in footnote 19, but that the total mass of the bodies of the

If we analyze this thought experiment *with the aid of the conceptual apparatus of relativistic dynamics*, we can describe this situation by saying that the *rest energy* of the system has decreased by an amount equal to m_0c^2. If we now *also* invoke the law of the conservation of energy (in the generalized sense in which this law transcends mechanics, both classical *and* relativistic), we would expect an increase in other forms of energy in the system by the same amount. We would expect, that is, that the sum of the kinetic, potential, thermal, . . . energies of the other bodies of the system would be increased by an amount equal to m_0c^2. (And conversely, if a mass m_0 suddenly appeared in the system, the new rest energy would be compensated by a decrease in other forms of energy for the other bodies in the system.)

It is this consequence of relativistic dynamics *plus* the law of the conservation of energy (in the generalized sense discussed earlier) that has been referred to as the *interconvertability of mass (or even "matter") and energy.*

Now since it is this consequence of relativistic dynamics which has seemed to some writers to be part of the revolutionary impact of relativistic dynamics, we must pause to see precisely what is being said here.

It is of the utmost importance to see that relativistic dynamics does *not* imply that mass can be converted into energy (and vice versa)! There are two reasons why this is so:

(1) The consequence in question requires not only the laws of relativistic dynamics but also the law of conservation of energy, and the latter is not part of mechanics alone.

---

system has decreased by an amount equal to the original mass of the body in question. (I am thus supposing that mass is not conserved.)

It is no more deducible within relativistic dynamics than it was within classical dynamics. (See pages 138–139.) In this sense, then, the introduction of *any* new form of energy is as "revolutionary" *per se* as the introduction of any other.

(2) More importantly, there is an essential "if" involved which most (if not all) writers seem to have missed. From relativistic dynamics (in particular, eq. (9)), plus the law of conservation of energy, one can deduce that *IF* a massive body should "disappear," *then* the previously recognized forms of energy of any closed system of which it was a part *would* increase by an amount equal to m_0c^2. But nowhere in relativistic dynamics will one find the statement or prediction that massive bodies *can* in fact be destroyed, with a consequent increase in energy in their environment. Nor will one find any deduction regarding the physical conditions under which such an event might possibly occur. From the standpoint of relativistic dynamics the actual occurrence of any such "annihilation of mass" would be a brute empirical fact, totally unexplained and unpredicted! In brief, *relativistic dynamics does NOT imply the interconvertibility of mass (or "matter") and energy*. What the relativistic theory *does* imply is an if-then statement telling us how the law of conservation of energy may be maintained in the event that we *independently discover* that mass can be made to disappear or appear.

Subsequent to Einstein's introduction of the concept of rest energy it *was* independently discovered that there are in fact situations in which massive bodies can "go out of existence." In these situations (in the domain of elementary particle physics), moreover, there *is* in fact an increase in other previously recognized forms of energy, and this increase *is* in fact equal to the lost mass multiplied by c^2. It is perhaps this verification of the if-then prediction of relativistic dynamics that has impressed those who speak of the revolutionary impact of the mass-energy relation. One might well imagine, after all, the mass-energy relation being discovered *experimentally* prior to Einstein's presentation of relativity theory, in the same way in which *thermal* energy was experimentally discovered before the advent of the kinetic theory of heat.

The revolutionary fact, if it can be deemed such, is that mass can in fact disappear, with ("other forms of") energy appearing. Relativistic dynamics plus the law of conservation of energy provides us with the proper bookkeeping rules for such occurrences, but these could have been developed experimentally, and (more importantly) the occurrences themselves are left unexplained by the theory. (An as-yet unavailable elementary particle physics—a "matter theory of the microcosm"—is needed for that.)

THE MEANING OF 'ENERGY'

I said earlier that I would not attempt here to decisively answer the question of whether the term 'energy' has a revolutionary *new meaning* in relativistic dynamics by virtue of the fact that the theory gives rise to a new form of energy "E," equal to m_0c^2.[21] (A full answer to this question, after all, would require an acceptable analysis of the meaning of 'meaning', including criteria for deciding whether or not a term has "changed in meaning." A discussion of this issue would not be appropriate here.)

But I would like to draw some conclusions from what I have said so far regarding the concept of energy. These conclusions, it seems to me, cannot be ignored if one wishes to settle the question of meaning change with respect to the

---

[21] Hence the famous equation $E = m_0c^2$.

concept of energy, as well as certain related philosophic issues.

First, and most importantly, we saw that the energy concept was introduced in relativistic dynamics in *exactly the same manner* in which it was introduced in the classical theory. In each case this was through the concept of work, defined as the integral over space of the force acting on a body. (See eq. (1), in *both* of its occurrences.) The only difference between the classical and relativistic theories in this respect is that different functions are used for the force **F**, since Newton's second law of motion, $\mathbf{F} = d/dt(m\mathbf{v})$, was replaced in the relativistic theory by

$$\mathbf{F} = \frac{d}{dt}\left(\frac{m_o\mathbf{v}}{\sqrt{1 - v^2/c^2}}\right).$$

I would suggest, without claiming to prove—indeed it is difficult to know what 'proof' means in this context, that on this basis it would seem more plausible to say that the term 'energy' has *not* changed in meaning. The opposite conclusion would follow only if it were assumed that 'F' had changed in meaning simply on the ground that it assumes a new functional form in the relativistic theory, and I find this ground insufficient.

Second, the energy concept proved important in both the classical and relativistic theories because of the general conservation law involving it. In this regard there is no difference between the classical and relativistic *use* of the concept. In *both* theories the law of conservation of energy serves the same function, and in each case the law transcends dynamics, involving as it does certain nonmechanical "forms of energy"—the same ones (thermal, and so on) for each theory.

There *is* this difference: The relativistic theory does give rise to a new form of energy: rest energy or ("mass energy").

I would suggest—again without claiming to prove—that this fact is insufficient for concluding that the *concept* of energy is therefore different, unless one is willing to maintain that each new form of energy really involves a change in meaning of the term 'energy'. But it sounds equally plausible, if not more so, to speak of the *discovery* of new *forms* of energy, where the *meaning* of the term 'energy' does not change (but is laid down once and for all through the concept of work and the general conservation law, for example).

Third, I would suggest that what is revolutionary is the fact that *mass* is now viewed as a form of energy. But it should be reiterated that relativistic dynamics does not *imply* that mass can in fact be *converted* into other forms of energy. As I said before, this fact is an independent discovery.

Finally, the revolutionary character of the fact that mass is not conserved after all does not by itself *imply* that the term 'energy' (or 'mass', for that matter) has changed in meaning.

To summarize, in my view there is not sufficient grounds for the claim that the term 'energy' has a different meaning in relativistic dynamics than it has in classical dynamics. Any successful argument to the contrary would have to take into account the considerations I have just discussed.

$E = mc^2$ AGAIN

I shall conclude this chapter by briefly discussing another facet of the mass-energy relation in relativistic dynamics.

The relativistic expression for work given in eq. (9) uses the symbol 'm_0' rather than 'm' for the concept of mass. (We could have used 'm_0' in place of all occurrences of 'm' in the earlier chapters on classical dynamics.) Let us refer to m_0 as the *rest mass* of a body, indicating that it refers to the mass of a body as determined by measurements in a frame of reference with respect to which the body is at rest. Rest mass is therefore a quantity intrinsic to a body, unlike (say) velocity.

We now reintroduce the symbol 'm' by the explicit *definition*

$$m = \frac{m_o}{\sqrt{1 - v^2/c^2}} \qquad \ldots (10)$$

and refer to m as *mass* (simpliciter). With this shift in terminology the *mass* (eq. (10)) of a body is a function of its velocity, and since velocity is relative to a frame of reference, mass (m, not m_0) is not a quantity intrinsic to a body but is relative to the frame of reference being used. The mass m of a body can take on any finite value between its rest mass m_0 and ∞.

We may now rewrite eq. (9) as

$$W = mc^2 - m_o c^2$$

or

$$W = (m - m_o)\, c^2 \qquad \ldots (11)$$

and say that when a force does work on a body, the effect of this is to *increase the mass of the body* from m_0 to m. What was previously described as an *energy increase* of amount $(m - m_0)c^2$ can *also* be described as a *mass increase* of amount $(m - m_0)$. This result—or, better, *redescription*—is sometimes referred to by saying that relativistic dynamics shows that *mass and energy are equivalent*.[22] (Notice that this is distinct from the *interconvertibility* of mass and energy discussed earlier. There is no question here of rest masses disappearing.)

---

[22] The equation $E = mc^2$ has also been used to express *this* result.

This *redescription,* via eq. (10), of energy increases as mass increases, has the advantage of allowing for the preservation of certain expressions from classical dynamics. Thus, relativistic *momentum*[23] can now be expressed as

$$\mathbf{p} = m\mathbf{v}$$

and the relativistic *force law*[24] can be expressed as

$$\mathbf{F} = \frac{d}{dt}(m\mathbf{v})$$

These equations are *identical* in form with their classical counterparts, the difference being hidden by the fact that '*m*' is not logically primitive as it was in the classical theory, but is defined as the product of the classical notion of mass, now referred to by 'm_0', and a certain function of velocity.

The "mass-energy equivalence" we have been discussing has also been referred to as the *inertia of energy*, since the concept of mass which enters into both the classical and relativistic force laws is that of *inertial mass*, resistance to changes of velocity.[25] The increase of energy of a body can be described as an increase in its *inertia*. But, as far as relativistic dynamics is concerned, it is only *mechanical* energy that can be so described. It is, once again, an *extrapolation* based on the general law of conservation of energy (of "all forms") to assert that other forms of energy as well have their "mass-equivalent." And it is a (perhaps revolutionary) fact that the resistance of a body to changes in its velocity does in fact depend on the quantity of other forms of energy (for example, thermal energy) characterizing the

[23] See Chapter 10, eq. (3).

[24] See Chapter 10, eq. (4).

[25] In earlier chapters we also used the symbol '*m*' to stand for the concept of *gravitational mass*. The fact that we could get away with the blurring of this distinction is highly significant, forming part of the access to the General Theory of Relativity.

body. (Thus, for example, if the thermal energy of a body increases by the amount T, we can also describe this as an increase in its mass by the amount T/c^2. The body will in fact behave as if its mass has increased by that amount.) Since these other forms of energy might not involve *velocity* in their functional form, it is quite clear that their equivalence to mass is *not* a consequence of relativistic dynamics, in which it is energy increases due to increases in *velocity* that are shown to be describable as mass increases.

One final consequence of this is worthy of mention. It now is apparent that the law of conservation of energy and the law of conservation of mass are, in fact, one and the same law. But, once again, this result too is not an implication of relativistic dynamics, but is a (well-founded) generalization from that theory together with the overarching law of conservation of energy.[26]

---

[26] It is this overarching character of the energy conservation law —its application to different theories with different domains—that is perhaps responsible for the naturalness in sometimes referring to it as a 'principle' rather than a 'law'.

12

POSTSCRIPT

In this final chapter I shall draw together and amplify some points which were made in passing earlier in the text. The items to be presented here could have been discussed at the appropriate places in earlier chapters, but were postponed to avoid the risk of breaking the thread of development at those places.

ANALYTICAL MECHANICS

It became apparent in Chapter 11 that the *law of conservation of energy* plays a key role in both classical and relativistic dynamics. This observation, plus the fact that this law also applies to nonmechanical domains,[1] should prompt one to ask whether classical and relativistic dynamics could be reformulated—and perhaps generalized—in such a way that the concept of *energy* would occur in the *postulates* of these theories rather than being introduced in the elaboration of the theory on the basis of the concept of *force*, as was done in Chapter 11. If this were possible, then perhaps the concept of force could be introduced as a *defined* notion on the basis of the energy concept; and the fundamental laws of motion ($\mathbf{F} = m_0\mathbf{a}$ in classical dynamics and $\mathbf{F} = d/dt$ $(m_0\mathbf{v}/\sqrt{1-v^2/c^2})$ in relativistic dynamics) would then be *theorems* in the reformulation. This kind of reformulation is

---

[1] There are, we saw, nonmechanical forms of energy. As I said in Chapter 11, the law of conservation of energy "transcends dynamics."

indeed possible. There are, in fact, several versions of it; they are the subject matter of what I referred to as *analytical mechanics* in Chapter 2, p. 29, and in Chapter 10, p. 129.

A reformulation of this kind has two advantages. The first (and least important) is that the foundations of mechanics can then dispense with the concept of force as a primitive concept. This is a goal which has always attracted those physicists and philosophers who have what I have called "positivist inclinations," going back (at least) as far as Hertz,[2] who tried to formulate mechanics without the force concept at all, and including some physicists who even today claim that $\mathbf{F} = m_0\mathbf{a}$ serves to explicitly *define* the force concept[3] rather than being a substantive law of nature.

The second advantage derives from the fact that the energy concept transcends dynamics. Thus, the reformulations[4] under consideration are *not merely* rearrangements of the statements of dynamics (either classical or relativistic) such that postulates and theorems reverse their roles, but they yield, rather, much more general physical schemes in which the elementary mechanical theories[5] appear as instances. (Indeed, analytical mechanics can be formulated on the basis of concepts even more general than that of *energy*.)

From the point of view of analytical mechanics, then, elementary classical and relativistic dynamics appear as two

[2] See Heinrich Hertz, *The Principles of Mechanics Presented in a New Form* (Dover, 1956; translated from the German edition of 1899).

[3] I suggested in Chapter 2 that this interpretation gives a misleading picture of the science of dynamics.

[4] Whether we use the singular or the plural here depends on whether one wishes to consider analytical mechanics as one theory with different versions or as several related theories.

[5] By 'elementary mechanics' (either classical or relativistic) I mean the dynamical theories presented in earlier chapters, with *force* as a basic concept and with general *force laws* as postulates.

distinct instances of the general scheme, each of which purports to correctly describe (the second successfully) one particular physical domain: massive bodies in motion under the influence of forces. Other instances of the same general scheme would then describe other physical domains, and in this way we can generate an "analytical thermodynamics," "analytical quantum mechanics," and so on.

I shall now very briefly sketch the structure of these reformulations without going into any of the mathematical details. Two important "theories of analytical mechanics" are due to Lagrange and Hamilton. In each of these theories a certain key concept is introduced (the "Lagrangian function," L, and the "Hamiltonian function," H, respectively), and a *developmental equation* is postulated[6] which states how the function (L or H) is related to the parameters of the system, including the time. Problem solving in specific situations amounts to stating the particular form for L or H for a particular type of physical system, and then solving the developmental equation with the given particular L or H for the coordinates as functions of time.

Notice the structural similarity to problem solving in elementary mechanics, as discussed in Chapters 2 and 3. In that case **F** played a role methodologically similar to L or H, and $\mathbf{F} = m_0\mathbf{a}$ played the role of what I called the 'developmental equation' in analytical (Lagrangian or Hamiltonian) mechanics. Just as there are *specific* L's or H's for specific physical systems or situations, there were specific force functions for these systems in elementary mechanics. And just as in elementary mechanics there was the inverse problem of determining the force function when the

---

[6] There is actually a *set* of such equations in each case, one for each "coordinate," where the concept of a coordinate is in a certain explicit sense more general than that which is used in elementary mechanics.

motion is given, in analytical mechanics there is the problem of determining L or H, given the observed temporal development of the "coordinates" of the system in question.

When these general schemes are applied to *mechanical* systems, the functions L and H are closely related to, but not always identical with, the expression in elementary mechanics for the *energy* of the system, and the Newtonian concept of force is introduced (defined) on the basis of the L or H functions. In this sense, then, elementary mechanics (both classical and relativistic) can be recovered from the Lagrangian and Hamiltonian versions of analytical mechanics as special cases.[7]

We can see now that the label 'analytical *mechanics*' is misleading. It is an *historical* fact that these theories grew out of attempts to "reformulate" Newtonian mechanics (hence Kuhn's "clarification by reformulation," quoted in Chapter 2), but they are *not* merely other versions of that theory. It would be better to refer to them by some domain-neutral term, such as 'general systems theory'. (Unfortunately this expression already has been adopted for other uses in cybernetics.)

The mathematics necessary for a more detailed development of analytical mechanics is beyond what I would like to assume of the reader, so I shall rest content with the outline I have just presented. But this has been sufficient to make one important negative point: What I referred to as the parasitical nature of analytical mechanics in footnote 7

---

[7] There is also a sense in which analytical mechanics is parasitical upon elementary mechanics. In concrete problem-solving cases of the inverse kind (find L or H when the observed temporal development of the coordinates is given) one attempts to solve the related problem in elementary mechanics *first*. The physicist will often construct the *force* and *energy* functions, and *then* use them to construct the Lagrangian L or the Hamiltonian H.

makes it evident that an understanding of the transition from classical to relativistic mechanics can be gained through presentations of the elementary versions of those theories, and does not require the more sophisticated elaborations of these theories which use the apparatus of analytical mechanics.

THE CONCEPT OF FORCE

When I presented elementary (classical and relativistic) dynamics in earlier chapters, I treated the concept of force as *primitive*. In this respect both force and mass (or rest mass m_0) are on a par (and both $\mathbf{F} = m_0\mathbf{a}$ and $\mathbf{F} = d/dt(m_0\mathbf{v}/\sqrt{1-v^2/c^2})$ are empirical postulates). For those who find themselves dissatisfied and want a "deeper explanation of what force really is" I would point out that the same question should arise (or else neither should arise) for the concept of mass. In each case we have concepts which are logically primitive in elementary dynamics. In a loose sense their "meanings" can be *clarified* by showing, for example, what sorts of entities have masses or exert forces, by becoming familiar with how forces and masses are *measured*,[8] and so on; but one cannot, within elementary (classical or relativistic) dynamics, *define* these concepts on the basis of anything more primitive.[9] In other words, forces and masses are not *reducible* (in the sense of Chapters 3

---

[8] If classical or relativistic dynamics were presented as formal axiomatic systems, then the provision of techniques for *measuring* forces and masses would not be stated within the syntax of these systems. They would be provided by (complex) "semantical rules" which relate these terms to the real world of gadgetry on a laboratory bench.

[9] Of course, for certain (limited) purposes in certain contexts one could treat $\mathbf{F}$ as defined by $m_0\mathbf{a}$, or in other contexts treat m_0 as defined by $\mathbf{F}/\mathbf{a}$.

and 4) to anything more physically basic, and $\mathbf{F} = m_0\mathbf{a}$ and its relativistic analogue are not deducible from any more fundamental mechanical laws. Within certain (speculative) versions of General Relativity Theory the situation is otherwise. There, both force and mass are interpreted as derivative notions, based on certain aspects of the structure of space-time.

I would suggest that those who seek a "deeper explanation" of force (and mass—or energy, for that matter) are asking for a *deeper physical theory* within which $\mathbf{F} = m_0\mathbf{a}$, or $\mathbf{F} = d/dt(m_0\mathbf{v}/\sqrt{1-v^2/c^2})$, would appear as *theorems*. They are not asking a question which can be answered within (either classical or relativistic) dynamics.

There is also the possibility that one who asks the "What is force (or mass)?" question is looking for a metaphysical gloss.[10] If so, one would have to know the metaphysical stance of the questioner before even attempting an answer.

THE RELATION BETWEEN NEWTON'S FIRST AND SECOND LAWS

In Chapter 2 (p. 15) it was pointed out that Newton's first law of motion—a body upon which no forces are acting will move uniformly in a straight line (that is, with constant velocity)—can be *derived* from the second law of motion, $\mathbf{F} = m_0\mathbf{a}$. For if $\mathbf{F} = 0$, then $m_0\mathbf{a} = 0$, or $\mathbf{v} = $ constant.

This fact gave rise to the question of why the first law is usually stated as a separate law. After all, the postulational

[10] For example, in Kant *force* is one of the pure derivative concepts of the understanding, one of the "predicables of the pure understanding." He speaks of "placing under the category of causality the predicables of force, action, passion" (*Critique of Pure Reason*, B 108; p. 115 of the Kemp-Smith edition). We could say that for Kant force is an activity of substance.

base of classical dynamics would seem to contain a redundancy if one of the "basic laws" can be deduced as a special case of another.

In Chapter 2 I postponed an answer to this question, saying that a full analysis of the notion of a *frame of reference* would be a prerequisite to a full answer.[11] Such an analysis was offered in Chapter 5.

In Chapter 5 eq. (4), I gave what I called the correct and complete expression of Newton's second law of motion:

$$(\phi) \left\{ P_\phi \equiv (\xi) \left[\mathbf{F}\,(\xi, \phi) = m_o\,(\xi, \phi)\,\mathrm{a}\,(\xi, \phi) \right] \right\}^{12}$$

We saw also that there appears to be no noncircular way, within classical dynamics, of defining the notion of a P-type frame of reference. That is, if we use the label 'inertial frame of reference' for those frames in which Newton's second law "holds," then in looking for some characterization, P, of such frames *independent* of Newton's second law, the best we can do within classical dynamics is refer to the brute fact that frames in uniform motion with respect to the distant galaxies are inertial frames—are P-type frames. (See pp. 72–73 of Chapter 5.)

But we can do a bit better than that, with the aid of Newton's *first* law, as I shall now attempt to show.

The very fact that Newton's first law is a special case of the second law can be used to characterize the notion of an inertial frame of reference. Thus we can say that an inertial frame, a P-type frame, is defined as one in which bodies upon which no forces are acting move with zero acceleration (constant velocity). Newton's *second* law of motion, as expressed in eq. (4), would then say (in English):

"If ϕ is any frame of reference in which bodies with *no*

[11] In Chapter 8 (p. 110) a partial answer was given.

[12] See the explanation in English of this symbolic expression in Chapter 5.

forces move with zero a (that is, if ϕ is an 'inertial frame'), then (and only then) in such a frame any massive body ξ with *nonzero* forces will move with *nonzero* acceleration a given by $\mathbf{F}(\xi, \phi) = m_0(\xi, \phi)\mathbf{a}(\xi, \phi)$."

In this way Newton's first law, *since* it is a (formally) special case of the second law, can be treated as *methodologically more basic*. We say, in essence: Consider those frames of reference with respect to which free bodies move uniformly. How will *un*free bodies move *in such frames*? (We put Newton's first "law" in place of '*P*' on the left side of eq. (4), and then ask what should be placed on the right side.) Once this question is answered (by $\mathbf{F} = m_0\mathbf{a}$),[13] it should be no surprise that in such frames the expression of Newton's first law will fall out as a special case.

Looked upon in this way, the relativistic force law, or its related mathematical equation $\mathbf{F} = d/dt(m_0\mathbf{v}/\sqrt{1-v^2/c^2})$, can now be viewed as a different answer to the *same question*. The first law thus appears as *methodologically basic* to *both* classical *and* relativistic dynamics, and will therefore appear, formally, as a special case of *both* dynamical theories.[14]

It is only if one pushes into the background the fact that the first law can be used to *define* the backdrop of reference frames against which theories of dynamics are to be formulated, and focuses upon the mathematical development of these theories, that one is tempted to view the first law as a

---

[13] Note again (see Chapter 5) that the *mathematical equation* '$\mathbf{F} = m_0\mathbf{a}$' is *not* to be confused with Newton's second law, which is given by eq. (4)!

[14] Compare Chapter 8, p. 110, where the concept of an inertial frame in *relativistic* dynamics was defined on the basis of Newton's first law of motion. We saw there that the first law is invariant under a *Lorentz* transformation between frames moving uniformly with respect to each other.

postulationally redundant special case of the general force law (using either $\mathbf{F} = m_0\mathbf{a}$ or $\mathbf{F} = d/dt(m_0\mathbf{v}/\sqrt{1-v^2/c^2})$).

The interpretation I have just suggested still does not circumvent the fact that at bottom neither classical nor relativistic dynamics contains a theoretical characterization of inertial frames. That is, the question "What is there about inertial frames (frames in which free bodies move uniformly) that makes them inertial?" is still unanswered. Compare Chapter 5, pp. 72–73, where the same question arose when $\mathbf{F} = m_0\mathbf{a}$ was used to define the concept of an inertial frame. The advantage of using the *first* law to define the concept of an inertial frame lies in the fact that the definition is neutral with respect to possible dynamical theories that can be erected on its basis.

In Chapter 1 I drew a distinction between kinematics and dynamics. Throughout the remainder of the book I have concentrated on the latter. This has been primarily a presentation of the methodological foundations of classical and relativistic *dynamics*. There was a brief sketch of a few results of relativistic *kinematics* in Chapter 9, but the burden of the discussion has been devoted to the dynamical theory that Einstein erected on the basis of the kinematics of the Lorentz transformation equations.

The reader who is familiar with the kind of popularization of "relativity theory" mentioned in the preface will perhaps be puzzled by the fact that he has not found in this book any mention of such familiar items as the relativity of simultaneity, light cones, world lines, space-time, the four-dimensional world, and so on. There are a number of reasons why I have omitted an extended discussion of these and other concepts and results of relativistic kinematics, and I shall conclude with a brief explanation of this omission.

First, it is my belief that a sound presentation of relativistic kinematics—one which would not run the risk of becoming a grab bag of strange and delightful ideas—requires a mathematical background which I have not wished to assume of the reader.[15]

Second, most popularizations and semiphilosophical presentations of relativity theory concentrate on the kinematical aspects and results of relativity theory, and tend to ignore or give inadequate attention to dynamics. I therefore thought it worthwhile to present here a sustained account of the foundations of Einstein's *dynamics*.

Related to this point is the fact that most of the literature on the philosophical implications of the "Special Theory of Relativity"[16] is devoted to relativistic *kinematics*, the new conceptions of space and time. But relativistic *dynamics* too is a rich field for philosophical analysis and speculation. If this book results in drawing more of the attention of philosophers of science to concepts such as *force*, *mass*, and *energy*, I will consider it a success.

[15] We would require a concept of a *vector* which is more sophisticated than that of a magnitude with a direction (an "arrow with three components"), for example. At bottom *tensor analysis* would be desirable.

(With such a background a more sophisticated presentation of relativistic *dynamics* would also be possible, one which leads in a natural way to the *General* Theory of Relativity.)

[16] See Chapter 10, p. 130.

APPENDIX A

Suppose we are given the motion of a mass attached to a spring, and the problem is to find the force responsible for the motion. From Newton's second law we have

$$F = m\ddot{s} \qquad \ldots (1)$$

and the equation for the observed motion

$$s = D \cos \omega t \qquad \ldots (2)$$

(D is the observed maximum displacement, and ω represents the frequency of oscillation.)

The problem is: Find some function f of the parameters of the situation which when substituted into the left side of eq. (1) will, upon integration, yield the observed motion expressed in eq. (2).

Differentiating eq. (2) twice, we obtain

$$\ddot{s} = -\omega^2 D \cos \omega t \qquad \ldots (3)$$

Putting eq. (3) into eq. (1), we obtain

$$F = -m\omega^2 D \cos \omega t \qquad \ldots (4)$$

Comparing eq. (4) with eq. (2), we note that the function

$$f = -m\omega^2 s \qquad \ldots (5)$$

will give back the observed motion when put into eq. (1). But is this a solution to the problem of finding the force law responsible for the observed motion? Do we now have a mechanical *explanation* of the oscillating motion of the mass? The answer is No, for the parameters in a satisfactory force law should show explicitly the way in which the nature of the spring—its stiffness—is responsible for the details of

165

the observed motion, and the appearance of 'm' in the force law obscures this. Suppose we set $\omega^2 = k/m$ in eq. (5) where k is a constant such that '$\omega^2 = k/m$' will be true. Then $\omega = \sqrt{k/m}$, and eq. (5) becomes

$$f = -m(k/m)s = -ks \qquad \dots (6)$$

This force law *is* satisfactory because k finds a *physical interpretation* as the stiffness coefficient of the spring, a constant depending solely on the nature of the spring. The force law expressed in eq. (6) satisfies our intuition that the force is proportional to the displacement, where the constant of proportionality is a measure of the stiffness of the spring.

We have restricted the set of parameters allowed to appear in a *physically* satisfactory force law. The implicit injunction or constraint in this example is: The force law should express the dependence of the force on two distinct factors: displacement and the nature of the spring. The mass m should not occur explicitly.

Consider also the case of the damped oscillator. We have once again Newton's second law, in eq. (1), and the equation for the observed motion:

$$s = De^{-\alpha t} \cos \beta t \qquad \dots (7)$$

Again, the problem is to find a function f of the parameters of the situation which when substituted into eq. (1) will, upon integration, yield eq. (7).

Differentiating eq. (7) twice, we obtain

$$\ddot{s} = -D(\beta^2 - \alpha^2)e^{-\alpha t} \cos \beta t \qquad \dots (8)$$

Putting eq. (8) into eq. (1), we get

$$F = -mD(\beta^2 - \alpha^2)e^{-\alpha t} \cos \beta t \qquad \dots (9)$$

Comparing eq. (9) with eq. (7), we note that the function

$$f = -m(\beta^2 - \alpha^2)s \qquad \dots (10)$$

will give back the observed motion when put into eq. (1).

But once again, it does not give a mechanical *explanation* of the damped oscillatory motion that is observed. The reason is the same as before. The parameters α and β cannot be physically interpreted as referring to the distinct aspects of the situation responsible for the motion. In this case there are two: the stiffness of the spring and the viscosity of the surrounding fluid. The *physically* satisfactory force function that we had in Chapter 2

$$f = -ks - c\,\dot{s}$$

contains coefficients which *are* so interpretable; and if we compare eq. (6) in Chapter 2 with eq. (7) above, we can see how we *did* have an explanation of the motion in Chapter 2, for the equation of the motion given there is in terms of k and c rather than the uninterpreted constants α and β.

APPENDIX B

In Chapter 4 (p. 47) Maxwell's equations for empty space were stated in what is known as their *differential form*:

$$\nabla \cdot \mathbf{E} = \rho \qquad \qquad \ldots (1)$$

$$\nabla \cdot \mathbf{H} = 0 \qquad \qquad \ldots (2)$$

$$\nabla \times \mathbf{E} = -\frac{1}{c}\frac{\partial \mathbf{H}}{\partial t} \qquad \qquad \ldots (3)$$

$$\nabla \times \mathbf{H} = \frac{1}{c}\left(\frac{\partial \mathbf{E}}{\partial t} + \rho \mathbf{v}(\rho)\right) \qquad \ldots (4)$$

They may also be expressed in their *integral form*, in which they are more amenable to "clarification," in the sense that in this latter form one can more easily obtain an intuitive grasp of what they say—how they relate the vector fields $\mathbf{E}$ and $\mathbf{H}$ and the scalar field ρ to one another.

THE FIRST TWO MAXWELL EQUATIONS

Consider an arbitrary region of space with volume V and surface area S. Now choose an arbitrary point on the surface and consider a "very small" element of area $d\sigma$

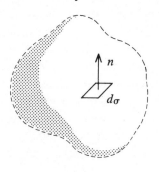

around this point. Within this area erect a vector, pointing outward, which is perpendicular to the surface and whose length is equal to one. This is the vector **n** shown in the figure above. It is called the "normal vector," and each element $d\sigma$ of the surface S will have such a vector associated with it. (The point of this construction will soon become apparent.)

We now integrate both sides of the first Maxwell equation over the region V to get

$$\iiint_V (\nabla \cdot \mathbf{E})\, dV = \iiint_V \rho\, dV \quad \ldots (5)$$

The right side of eq. (5) is just the "sum" of all the charges inside the region, e_V. That is,

$$\iiint_V \rho\, dV = e_V$$

Also, there is a theorem in vector analysis (called the "divergence theorem") which states that for any vector field **Q**

$$\iiint_V (\nabla \cdot \mathbf{Q})\, dV = \iint_S (\mathbf{Q} \cdot \mathbf{n})\, d\sigma$$

Equation (5) can therefore be rewritten as

$$\iint_S (\mathbf{E} \cdot \mathbf{n})\, d\sigma = e_V \quad \ldots (6)$$

This is the integral form of $\nabla \cdot \mathbf{E} = \rho$. '**E** · **n**', called the "dot product" of the two vectors, is a *scalar* quantity defined as the product of the magnitude of **E**, the magnitude of **n** ($= 1$), and the cosine of the angle between them. It is a measure of the extent to which the vector field **E** is crossing the surface of the region in the vicinity of the element of area $d\sigma$ upon which the vector **n** was erected. (Thus, if the angle between **E** and **n** at that element of area is zero, its cosine is also zero, and the electric field vector **E** is tangent to the surface; the field is not "passing through" the surface at the chosen element of area.) The sum of all these "cross-ings-over" of **E** is the integral on the left side of eq. (6), and is called the "total flux of **E** through *S*." Eq. (6), then, states that *the total flux of* **E** *through an arbitrary closed surface is equal to the total electric charge contained inside the surface.*

If this quantity is positive, there is a net *positive charge* in *V*; we say that there are *sources* of the field in *V*. If it is negative, there is a net *negative charge* in *V*; we say there are *sinks* of the field in *V*. If we allow our arbitrary region to shrink to an arbitrarily small "closeness" to any chosen point within it, we can recover the original differential form of the first Maxwell equation. Eq. (1) states that *the divergence of* **E**, *(∇ · E), at a point is a measure of the charge density at that point.* (The divergence of a vector, at a point, is a measure of the extent to which the vector field "radiates" from that point. Thus we may picture electric charges as having arrows representing the **E**-field radiating from them, pointing outward for positive charges and inward for negative charges.)

The integral form for the second Maxwell equation, $\nabla \cdot \mathbf{H} = 0$, will be

$$\iint_S (\mathbf{H} \cdot \mathbf{n})\, d\sigma = 0 \qquad \ldots (7)$$

We see now that this can be interpreted as stating that *there are no magnetic charges.* The magnetic field vector **H** has no sources or sinks.

THE SECOND TWO MAXWELL EQUATIONS

The first two Maxwell equations related the field vectors to their sources. Electric fields are related to (produced by) electric charges. Magnetic fields have no sources of their own; they owe their existence to electric fields, as indicated by the second two Maxwell equations, which relate the fields to each other. These equations can also be written in integral form with the aid of constructions similar to those used above, together with another theorem from vector analysis. I shall not do so here. It is sufficient for our purposes to note that eq. (3) indicates that a changing magnetic field will give rise to an electric field, and eq. (4) indicates that a changing electric field will give rise to a magnetic field. (Eq. (4) also states, via the second term of the right-hand side, that moving electric charges will also give rise to a magnetic field.) Most of these facts were learned experimentally before Maxwell developed his equations, from which they can be deduced.

One final word. I said above that electric fields are produced by electric charges. That is, if charges are present, then electric fields will be present. But Maxwell's equations do *not* imply that if an electric field is present, then charges must be present also. For Maxwell's equations allow for the possibility of infinitely extended uniform, electric fields with no electric charges. It is an independent fact (or well-grounded belief) that there are no such fields in the universe: that electric fields are present if *and only if* electric charges are present.

INDEX